Praise for *The Lean Product Lifecycle*

The world is becoming a highly strung, highly digital marketplace where agility and efficiency is key. Businesses need to adapt traditional methodology to ensure they can maintain competitiveness. *The Lean Product Lifecycle* is an invaluable guide to help you future proof your company, ensuring innovation is at the forefront whilst managing the core business.

**Antonia Barton, Marketing and
Digital Director, BT Plc**

How can corporations innovate faster and better? *The Lean Product Lifecycle* provides a really practical way to do this. The book delivers a clear roadmap for developing and managing products along their lifecycle. This book is a must-have practical guide for corporate innovators.

Thomas Krogh Jensen, CEO, Copenhagen Fintech

The Lean Product Lifecycle is crammed full of practical tools and examples for bringing lean innovation into any company. A must-read.

Des Dearlove and Stuart Crainer, founders, Thinkers50

The path to profitable new products that has invigorated our product development process.

**Peter Pascale, Vice President,
Product Management, Pearson VUE**

The product life cycle model is a key tool for product people. This book does a great job at showing how the model can be applied in a lean way to help companies innovate faster.

Roman Pichler, product management expert

Corporate innovation is becoming more and more important. Tendayi, Craig and Sonja have done a wonderful job detailing a lean innovation toolbox that can be used to help innovators get to their goals, faster.

Ole Madsen, Senior Vice President,
Spar Nord bank

Most companies dream about change and innovation but few know how to make it happen. *The Lean Product Lifecycle* team has the vision, experience and battle scars to help you kick start a meaningful transformation in the way you engage with customers, manage your product portfolio and make better business decisions. I highly recommend it to any corporate change agents looking for a great weapon in their arsenal.

Juan Lopez-Valcarcel, former Chief Digital Officer,
Pearson International

If your business is about winning with technology and product innovation, then this is the ideal book to help guide you. *The Lean Product Lifecycle* sets the stage for agility and a very smart approach to streamlining your competitive edge! Easy to read and laced with real world interviews and observations that give context and currency to what this skilled team of authors are sharing with you.

Phil Blades, experienced exec and advisor to high-tech
start-ups and global enterprises

You've made it. You have a tech company, you have customers, and a revenue stream. But how do you sustain your success, continue to grow and remain competitive? You need to innovate. *The Lean Product Lifecycle* is your guidebook for sustained innovation. Through their combined product development and go-to market experience, the writing team shares industry best practices that work, and the tools to accelerate your success. *The Lean Product Lifecycle* is a must have for management and product teams who make innovation a priority.

Angus Robertson, CMO, Axcient

The Lean Product Lifecycle

Pearson

The Lean Product Lifecycle

How to make products people want

Tendayi Viki
Craig Strong
and Sonja Kresojevic

 **Pearson**

Harlow, England • London • New York • Boston • San Francisco • Toronto • Sydney • Dubai • Singapore • Hong Kong Tokyo • Seoul • Taipei • New Delhi • Cape Town • São Paulo • Mexico City • Madrid • Amsterdam • Munich • Paris • Milan

Pearson Education Limited
KAO Two
KAO Park
Harlow
CM17 9NA
United Kingdom
Tel: +44 (0)1279 623623
Web: www.pearson.com/uk

First edition published 2018 (print and electronic)

ISBN: 978-1-292-18641-2 (print)
         978-1-292-18689-4 (PDF)
         978-1-292-18690-0 (ePub)

British Library Cataloguing-in-Publication Data
A catalogue record for the print edition is available from the British Library

Library of Congress Cataloging-in-Publication Data
A catalog record for the print edition is available from the Library of Congress

ARP impression 98

Cover design by Madras
Print edition typeset in 9.5/15 pt Open Sans by Pearson CSC
Printed by Ashford Colour Press Ltd

NOTE THAT ANY PAGE CROSS REFERENCES REFER TO THE PRINT EDITION

Do you manage or deliver products within a startup or an established company?

Yes_____ No_____

Have you been looking for ways to apply lean innovation methods to your practical day-to-day work?

Yes_____ No_____

Are you a leader wanting to spark innovation within your company?

Yes_____ No_____

Have you been looking for innovative ways to invest and manage products across your portfolio?

Yes_____ No_____

If your answer was 'Yes' to one or more of these questions, then this is the right playbook for you. With the world changing rapidly, the ability to innovate and sustain growth is now the only competitive advantage a company can have. The Lean Product Lifecycle (a.k.a. Lean PLC) is for game changers looking to spark sustained innovation and growth within their companies. A game changer can be anybody: factory worker, product manager, product owner, designer, software developer, head of innovation, head of digital, head of sales and marketing, head of finance, senior managers and C-Level executives. Our goal is to provide you with an actionable toolbox that will enable you to do the right things at the right time. So welcome to The Lean Product Lifecycle: A Playbook for Making Products People Want. We hope you enjoy the journey.

Contents

About the authors

Tendayi Viki is a highly experienced innovation leader. He helps large organisations develop their ecosystems so that they can innovate for the future while managing their core business. He holds a PhD in Psychology and an MBA. Through his company Benneli Jacobs, he has given keynotes, run workshops and worked as a consultant for several large organizations including Rabobank, American Express, Standard Bank, Unilever, Airbus, Pearson, Lufthansa-Airplus, The British Museum, Copenhagen Fintech and The Royal Academy of Engineers. Tendayi has been shortlisted for the Thinkers50 Innovation Award and was named on the Thinkers50 2018 Radar List for emerging management thinkers to watch. He is also a regular contributing writer for Forbes. Tendayi spent over 12 years in academia during which time he taught at the University of Kent where he is now Honorary Senior Lecturer. He has also been a Research Fellow at Stanford University and Research Assistant at Harvard University.

Craig Strong is an experienced Chief Technology and Product Officer, specialising in growth and innovation, who has grown companies to successful exits and acquisitions. He is also a member of Forbes Technology Council and was one of the founding members of Agile Practitioners London. For over 18 years, Craig has been developing and delivering market leading software and technology solutions focused on customer needs, helping companies and teams innovate, scale and grow. Heavily involved with global enterprise transformation and innovation programmes utilising Agile and Lean principles with an adaptive management approach, he has worked and consulted for companies which include Sky, NowTV.com, Hargreaves Lansdown, Pearson, Financial Times, InsightSoftware.com, Tracesmart and others.

Sonja Kresojevic is a senior executive with 20 years of global experience transforming businesses by focusing on growth, product innovation, cultural change and digital transformation. She is a co-founder of the NY based consultancy Spinnaker Consulting, helping Fortune 500 organisations discover new growth opportunities, deliver great products and become more innovative under conditions of extreme uncertainty and disruption. In her most recent corporate role, she was a Senior Vice President in the Chief Product Office at Pearson, leading a global team responsible for design and implementation of the Lean Product Lifecycle, award-winning innovation program focused on transforming product development and portfolio Investment management and delivering a faster and more innovative organisation. She is a speaker on topics of transformation, business agility, leadership, innovation strategy and the co-author of the upcoming book: *The Case for Change – Demystifying Lean Enterprise Transformation*.

Publisher's acknowledgements

We are grateful to the following for permission to use copyright material:

Chapter 1 interview with Sebastian Cadell, Partner at Nosco on page 19–21, reprinted by permission of Sebastian Cadell; Chapter 2 interview with Nikitas Magel, Senior Content Manager at Brightidea on page 42 and 43, reprinted by permission of Nikitas Magel; Chapter 3 interview with The Innovaid Team at Rabobank on page 90 and 91, reprinted by permission of Rabobank; Chapter 4 interview with Teodora Berkova, Director of Social Innovation, Pearson on page 119–121, reprinted by permission of Teodora Berkova, Pearson Education, Inc.; Chapter 5 interview with Marc Abraham, Chief Product Officer at Settled on page 169–171, reprinted by permission of Marc Abraham, Settled; Chapter 8 SDL - Rolling Out The Lean Product Lifecycle on page 226–230, reprinted by permission of SDL; Insights Software on page 230–235, reprinted by permission of Insights Software.

The Lean Product Lifecycle team

The business of education is changing rapidly. What was once a traditional publishing industry is being changed by digital technologies. There has been an increase in the number of EdTech startups that are trying to disrupt education. This shifting landscape puts pressure on incumbent companies that have dominated this industry for decades. One of those companies is Pearson, a FTSE 100 global education company with over 35,000 employees.

The Lean Product Lifecycle, was a response to a challenge that was set by Pearson for our team to develop a methodology for product development and investment decision making based on lean principles. The original Lean PLC team had five founding members: the three authors of this book, Elin Cathcart and Shirley Chin. The team was led by Sonja Kresojevic who was then Senior Vice President of Product Lifecycle. She brought her intelligence, energy and tenacity to the difficult work of transforming a large company; and we could not have succeeded without her leadership.

Over time the Lean PLC Team grew, changed and evolved. We have enjoyed working with wonderful colleagues who have been key contributors in the development and success of the framework. We would like to acknowledge and thank these team members for their key contributions: Adam Berk, Jonathan Bertfield, Stacey Dobbs, Sophie Freiermuth, Gabe Gloege, Jacqueline Krain, Chris Locke, Shani Malloy, Shannon Rendel and Julia Shalet.

The Lean PLC won the award for the *Best Innovation Programme 2015* at the Corporate Entrepreneur Awards in New York. We were also awarded the *Best Innovation Culture 2015* at the Corporate Entrepreneur Awards in London. We were the first global innovation team to win this award in both the American and European competitions; an achievement of which we are immensely proud.

On this journey, we have also worked with great partners within Pearson who have helped our team develop the framework. We would like to thank and acknowledge their contributions: Luyen Chou, Dan Ginsberg, Jason Goodhand, Alan Read, Freya Thomas-Monk, Jim Hummer, Kimberly O'Malley, Frances Soul, Sue Ann Averitte, Els Howard, James Kenwood, Amar Kumar, Susan Rudolph, Marc Bishara, Vivek Govil,

Diana Stepner, Nina Angelo, Dana Rodericks, Charlotte Baldwin, Cathy O'Brien, Tom Hall, Peter Pascale, Barbara Wilson, Andy Sales and Kelwyn Looi.

Our work was also influenced by wonderful colleagues outside of Pearson whose great work inspired us: Eric Ries, Alexander Osterwalder, Steve Blank, Tristan Kromer, Barry O'Reilly, Jeff Gothelf, Josh Seiden, Ben Yoskovitz, Farrah Bostic, Stuart Eccles, Marc Abraham, Roman Pichler, Paul Michael, Sam Whiting, David Leach, Geert Wirtjes, Kevin Josling, Naill McLean, Chelsea Rider and others.

The faces of the Lean Product Lifecycle

Elin Cathcart

Adam Berk

Sophie Freiermuth

Shirley Chin

Julia Shalet

Gabe Gloege

Jacqueline Krain

Stacey Dobbs

Shani Malloy

Jonathan Bertfield

Shannon Rendell

Chris Locke

Part 1

1

Introduction

1.1 A changing business landscape

To say the world around us has changed is to state the obvious. The world is always changing. What is remarkable is not the change *per se*.

Instead, what is remarkable is the *pace* of that change which has increased dramatically. What is also remarkable is the *extent* of the change, which has resulted in a significant paradigm shifts for business. Over the last ten years, we have seen the rise of several great startups and products that have leveraged technology to create new business models, disrupt incumbents and grow rapidly. It is remarkable to think that just ten years ago transformational companies such as Dropbox, Uber and Airbnb didn't exist.

In response to this rapid pace of change in the business environment, most incumbent companies have been slow to respond. In as much as we can list the great companies that have emerged in the last decade, we can also list examples of formerly great companies that have been disrupted. Examples include Kodak, Blockbuster, Nokia, HMV, Borders, BHS and Blackberry. These companies and others have found themselves in trouble owing to the impact of innovative technologies on their industries.

Most established companies found their success during a time when the pace of change was much slower. So for them, great strategy meant finding a competitive advantage and protecting it through good financial management.[1] These businesses were run on the basis of 3–5 year business planning, annual budgets, waterfall product development methods, siloed business departments (e.g. HR, marketing, finance and legal) and investment processes that relied on making big bets upfront (e.g. for product development or mergers/acquisitions). So just when they need to be more agile and responsive, these companies find themselves with management systems and processes that were most useful during a different era.

Nowadays, the notion of a stable business environment is a myth. Companies have to be managed to consistently move from one competitive advantage to another.[2] Good strategy is now about *exploiting* current advantages while *exploring* new ones.

---

[1] Ireland, R.D., Hoskisson, R.E. and Hitt, M.A. (2011). *The management of strategy.* London: Cengage Learning.

[2] McGrath, R.G. (2013). *The end of competitive advantage.* Harvard Business Review Press: Cambridge.

In such a world, companies need new management frameworks and ways of working. This book presents one such a framework. Based on design thinking, lean and agile methodologies, the Lean Product Lifecycle (Lean PLC) is a framework that helps companies make products people want and manage investment risk by making the right decisions at the right time.

Our framework provides a systematic way for companies to incrementally invest in new ideas, manage new product development, scale or sustain mature products and retire declining ones. Our goal is to provide companies with the right lens for examining their portfolio of products in order to make the right investment and product development decisions. The Lean PLC does not require established companies to become like startups; rather we provide tools and methods that allow enterprises to understand and apply lean thinking.

1.2 Lean thinking

Over the last 20 years, lean and agile methods have transformed the way products are developed and managed. Although technology startups have been at the forefront of developing and advocating lean and agile, these methods were not developed for startups only. Lean and agile are based on a set of underlying principles that, if well understood, can be applied to any company. Understanding and applying these principles is what can help transform our companies.

Lean thinking is not about being cheap or small. The absolute size of an investment is *not* what defines a process as being lean. Instead, lean thinking is the discipline of doing the right things at the right time. What is defined as waste in lean thinking is when companies do the wrong things at the wrong time. For example, one of the reasons why new products fail is *premature scaling* (i.e. spending too much money or resources building our products without testing with customers and then just launching those products to the broader market).

The practice of lean thinking is particularly challenging for established companies that are already operating at scale or startups that are keen to quickly build their product and release it to the wider market. How do they ensure that they are doing the right things at the right time? At the heart of the lean approach is the distinction between *searching* and *executing*. According to Blank and Dorf a startup is an organisation whose primary goal is to search for a sustainably profitable business model. In contrast, an established company spends most of its time executing on a known business model.[3]

Blank and Dorf use this distinction between startups and established companies to good effect here. But this focus on distinguishing startups from established companies can sometimes be misleading. What is really important here are not the different types of companies but the different types of products they are managing at any given time. Startups are usually working on new products that are yet to be proven in the market. They are still looking to understand customer needs, create the right solutions and find profitable business models. In contrast, most established companies have mature products that are already serving known customer needs, with proven business models.

Using the product perspective can help established companies adopt lean thinking. Rather than viewing their company as single institution with one business model, management can take a portfolio view. When they start thinking of their products as a portfolio, management can then examine each product and classify it as either in *search* mode or *execution* mode. Once they know which mode a product is in, this can then help them make more informed decisions about what to do next (i.e. the right thing at the right time).

The portfolio approach also helps companies view strategy as a way of *executing* on current competitive advantages, while *searching* for new ones.[4] So beyond managing their core cash-cow products, every company should also be working on future-facing product ideas. If the majority of a company's portfolio are mature products, this can be a signal for management to do something to redress this imbalance. Since disruptive innovations have affected every industry, having a balanced portfolio is an imperative for established companies.

---

[3] Blank, S. and Dorf, B. (2012). *The startup owner's manual.* K&S Ranch: California.

[4] McGrath, R.G. (2013). *The end of competitive advantage: how to keep your strategy moving as fast as your business.* Harvard Business Review Press: Cambridge.

1.3 Managing your portfolio

In order to achieve a balanced portfolio, companies can use frameworks that allow them to distinguish between different types of innovation. One framework we found useful in our work is the *innovation ambition matrix*. In an article for *Harvard Business Review*,[5] Bansi Nagji and Geoff Tuff used two main dimensions, *products* (new products *versus* existing products) and *markets* (new markets *versus* existing markets), to distinguish core, adjacent and transformational innovation.

- *Core innovation* involves making incremental changes to existing products for existing customers. Such innovation uses assets the company already owns and markets in which it is already present.

- A*djacent innovation* involves taking currently successful products to new markets or developing new products for current markets. The company is using existing capabilities but putting them to new uses.

- *Transformational innovation* involves creating new offerings for new markets. Here, the company is developing new products or services for new markets that it is currently not operating in.[6]

The strategic role of the innovation ambition matrix is to make sure that every company has a balanced portfolio of products and business models. It is important for every company to have within its portfolio, products and services that cover the range from core, adjacent and transformational innovation. Nagji and Tuff propose a magic ratio for companies to allocate resources within a balanced portfolio; i.e. 70 per cent in core, 20 per cent in adjacent and 10 per cent in transformational.

This ratio is not fixed and can be adapted depending on company and industry. The goal is for each company to make explicit its innovation ambitions, and then work towards balancing its investments accordingly. It is also important to make clear that the ratio is not about the exact number of the different types of products within a

---

5   Nagji, B. and Tuff, G. (2012). Managing your innovation portfolio. *Harvard Business Review, 90*(5), 66–74.
6   Nagji and Tuff's framework in similar to McKinsey's three horizons model. Baghai, M., Coley, S. and White, D. (1999). *The alchemy of growth: practical insights for building the enduring enterprise.* London: Orion Business.

company. It is about resources and how they are allocated across the different types of innovation.

In fact, a relatively small investment with just 5 per cent of company resources in transformational innovation can result in a larger number of new product ideas compared to core products, if the company uses incremental investment methods and innovation accounting. More on that later! We will now turn our attention to how companies should be managing the range of products within their portfolio, by doing the right things at the right time.

Figure 1.1

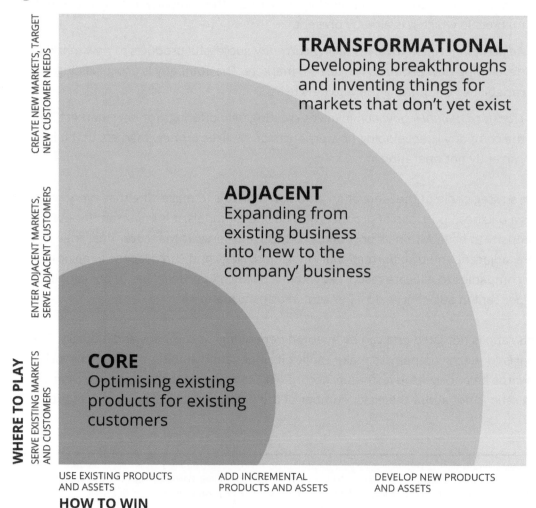

TRANSFORMATIONAL
Developing breakthroughs and inventing things for markets that don't yet exist

ADJACENT
Expanding from existing business into 'new to the company' business

CORE
Optimising existing products for existing customers

CREATE NEW MARKETS, TARGET NEW CUSTOMER NEEDS

ENTER ADJACENT MARKETS, SERVE ADJACENT CUSTOMERS

SERVE EXISTING MARKETS AND CUSTOMERS

WHERE TO PLAY

USE EXISTING PRODUCTS AND ASSETS

ADD INCREMENTAL PRODUCTS AND ASSETS

DEVELOP NEW PRODUCTS AND ASSETS

HOW TO WIN

1.4 Right thing, right time

It is important for us to recognise that the tools and methods for managing innovation are different from those that have traditionally been used to manage core products. The clue to this difference lies in the distinction we have already highlighted between *searching* and *executing*. Products that are still searching for sustainable business models should not be managed as if they are already at scale in the market. However, within that general distinction between searching and executing, there are key steps that companies need to follow.

To help articulate these steps, we first need a shared definition of what represents successful innovation. From our perspective, *successful innovation is the combination of creative new ideas with sustainable business models.*[7] A sustainable business model is made up of two parts: (1) products that serve real customer needs and (2) the ability to create and deliver those products to customers in a repeatable and profitable way. This definition of innovation allows us to break *searching* into three key steps: *ideation* during which we generate creative new ideas, *exploration* during which we test whether our ideas serve real customer needs and *validation* during which we create the solution and develop a profitable business model.

Once a sustainably profitable business model has been discovered, we have achieved so-called *product–market fit.* Only then are we ready to begin *executing* and taking the product to scale. In the early days of scaling, success is measured by how well we can *grow* customer numbers, revenues and profits. Over time, the product will reach its maturity and growth will slow down. At the point, the focus will move to *sustaining* revenues while optimising costs. As time moves on, the product starts to decline significantly it can be *retired* and moved out of our portfolio.

It is important to note that waste can occur during execution as well (e.g. adding features to product that don't add value to customers). As such, lean thinking applies across the whole lifecycle of a product. Doing the right thing at the right time means recognising the lifecycle stage that the product is in and using the correct product development methods and investment decisions for that stage. This is now *Management 101* for contemporary companies.

---

[7] Viki, T., Toma, D. and Gons, E. (2017). *The corporate startup.* Amsterdam: Vakmedianet.

Figure 1.2

1.5 How the Lean Product Lifecycle works

This idea of a product lifecycle is not new. The concept was first developed by Raymond Vernon in 1966.[8] Vernon saw products going through four stages; *introduction, growth, maturity* and *decline.* In this model, there is no requirement for innovators to search for a profitable business model before moving onto the growth stage. The Lean PLC was design to deal with this key shortcoming.

Our framework has six stages. The first three stages are on the left side: *idea, explore, validate.* This side of the Lean PLC is focused on searching for a profitable business model with the ultimate goal of getting to product–market fit (i.e. the dotted line). On the right side of the Lean PLC are the last three stages: *grow, sustain, retire.* This side of

---

[8]   Vernon, R. (1966). International investment and international trade in the product cycle. *The Quarterly Journal of Economics.*

the Lean PLC focuses on executing on a known business model, exploiting it with efficiency until the product is ready to be taken of the market.

We will now briefly describe each Lean PLC stage from the product development perspective. To really understand the practices that underlie the Lean PLC, it is always key to remember that *no product or service should be taken to scale in the market before its business model has been tested and validated.*

 Idea: During this stage, teams use customer and market insights to generate ideas. After coming up with a number of ideas, they should then choose one idea to work on, review that idea for risky assumptions and brainstorm ways to test these assumptions.

 Explore: During this stage, teams get out of the building and test their risky assumptions. The main focus of Explore is to test assumptions about customer needs, problems and job-to-be-done.

 Validate: During this stage, team use the insights from Explore to develop a solution for customers, starting with a minimum viable version. The point is to iteratively develop a product that delivers value to customers. The solution development process is also used to test aspects of our business model including pricing, costs and channels.

 Grow: A key consideration for entering the Grow stage is the traction we demonstrate during Validate. With a validated business model, now it's time to take our product to scale. The main focus during Grow is on increasing customer numbers, revenues and profits.

 Sustain: Over time, all products reach a level of maturity as markets get saturated, competitors enter the market or the technology changes. When this happens, the products enter the Sustain stage, during which the focus shifts to sustaining revenues and profits, while optimising operations and reducing costs.

Retire: Nothing last forever! Eventually, there comes a time when a product must be retired. During this stage, it is important to ensure that customers are not inconvenienced by putting in place a plan to mitigate negative effects on customers.

These are the product development best practices connected to the six stages of the Lean PLC. However, for teams to engage in such practices they need access to the right resources at the right time. As such, connected to product development are the ongoing investment decisions that are made by management. How managers make decisions to invest in or manage projects will have an impact on how well teams can use the Lean PLC to develop their products.

1.6 Spend less upfront

Premature scaling does not happen by itself. It is a problem that is fuelled by how companies have traditionally invested in new ideas. At the heart of every investment decision they make is the business case or plan. No new product is ever developed until the product team creates a plan that demonstrates a good return on investment (ROI), projected over three to five years. At the heart of the business plan is a request for an investment amount. This number is usually quite large, because it includes all aspects of developing, marketing and scaling the product.

We have seen companies invest millions in untested ideas on the basis of the business case alone. This is premature scaling, fuelled by corporate resources. Business planning works well for established products with a known trading history upon which to base our five-year projections. And even then, we have to be careful of potential shifts in our business environment. But when it comes to new ideas, especially transformational innovation, business planning is an exercise in fiction writing. When a plan is created that early in the process, most of what is in the plan are assumptions; even if they are written in the document as facts.

Until recently, there has not been any other management framework beside business planning that managers could use to make investment decisions. In 2010, Dave McClure published an article entitled *Moneyball for Startups.*[9] In the article, he outlined an investment process that involves making small initial investments in ideas while

Figure 1.3

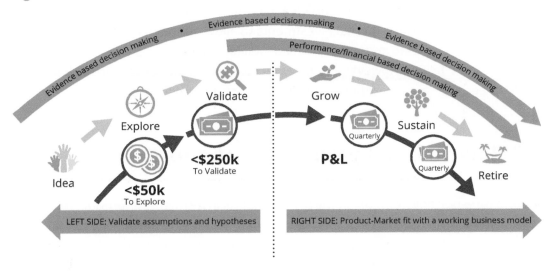

teams are searching for product–market fit. If the teams succeed at searching, investments in their ideas can then be increased. In practice, the small initial investments are used to test whether the product idea has potential before a lot more money is spent.

The Lean PLC provides a framework that can form the basis for established companies to play moneyball. During Idea, Explore and Validate the focus is mostly on testing the market, potential solutions and the business model. As such, small amounts of money can be invested and as ideas start to show traction, managers can then make larger investments. Each company can set its own upper limits for investment depending on industry and types of products. At Pearson, we recommended investing no money for the Idea stage, less than £50,000 GBP during Explore and less than £250,000 GBP during Validate.

When a product crosses to the right side of the Lean PLC, investments can then be based on P&L. This would be a validated P&L and any projections made at this stage would be based on learnings from the first three stages. This approach

---

[9] McClure, D. (2010). *Moneyball for startups: Invest BEFORE product/market fit, double-down AFTER.* 500 Hats. Available at http://500hats.typepad.com/500blogs/2010/07/moneyball-for-startups.html

chimes with our message to do the right things at the right time. There is a right time to make a large investment in an idea. It is after it has demonstrated traction. Before that, we make incremental investments to allow our teams to test ideas in the market. This is not to argue that no ideas will fail. Innovation is about accepting some failure. However, incremental investing allows us to learn quickly if an idea has legs and stop investments in failing projects before too much money is spent.

1.7 How to build your product council

So, who is responsible for making these incremental investments in products? It is possible for individual leaders or managers with the budget responsibility to make such decisions. At Pearson, we pioneered the concept of product councils. These are investment boards that are empowered to make decisions across the Lean PLC. Decision making by committee is sometimes frowned upon but we believe that a small cross-functional board can be useful in keeping individual decision makers accountable in terms of asking the right questions at the right time.

It is normal practice within companies that teams requesting funding have to prepare some sort of business case. Most companies use business case templates with the key sections that they expect each team to complete before their funding request is considered. The Lean PLC provides a framework for creating similar templates. The key difference is that we do not recommend using one generic business case template regardless of the lifecycle stage of product. Rather, we believe that templates should be based on the lifecycle stage and ask only the relevant questions for that stage.

In each chapter of this book, we will provide examples of templates that can be used by teams to make submissions and investment boards to make decisions based on the Lean PLC stage. A map of how each template aligns to a PLC stage is presented below. We will recommend specific questions and criteria that can be used to move a product from one stage to another. These criteria also inform the activity we recommend that product teams engage in during any particular stage. This alignment between teams doing the right things as they develop their products and managers asking the right questions before they make investment decisions reveals the power of the Lean PLC.

Figure 1.4

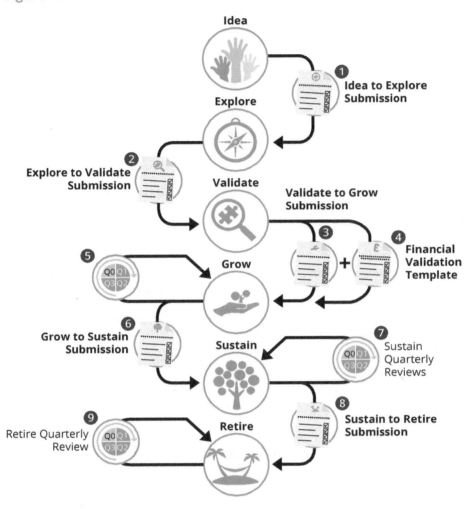

1.8 Not a straight line

The challenge with frameworks is that they are often mistakenly treated as canonical truths. Critics rightly point out that frameworks do not fully represent reality. This is something we fully agree with regards to our own framework. It is highly unlikely that any product team will move through the Lean PLC from stage to stage in a linear fashion. It is also unlikely that product teams and managers will find clear cut boundaries when trying to move a product from stage to stage. Instead, boundaries will be fuzzy and decisions will have to be judgement calls. This is the nature of innovation and business in general. It is a nonlinear mess that can best be described as organised chaos.

So why have we created a framework at all? And why have we presented it in linear fashion? Luke Hohmann describes frameworks as tools for knowledge workers.[10] Given the non-concrete nature of knowledge work, concepts around innovation and management can be harder to bring to life. Frameworks provide reference points that allow knowledge workers to come to grips with complex and ambiguous situations. The Lean PLC takes principles and practices from multiple disciplines and combines them into a usable framework.

That being said, it is important to recognise that the Lean PLC is described in a linear fashion for story-telling purposes only. This lifecycle story makes it easy for people to understand how the Lean PLC works. In practice, however, the Lean PLC is a nonlinear. Not all products in a company will enter the Lean PLC at the idea stage. In every company we have worked with, there has been a portfolio of products that were at different lifecycle stages. Each product enters the PLC at the appropriate stage. What we provide are a series of questions and analytical tools to help businesses classify their products.

The journey of a product along the Lean PLC is also nonlinear. Products do not move between stages in one direction only. It is possible for product to skip a stage or move back a stage. For example, a team can get strong signals around customer needs. However, when they create the solution, they then find that customers are not as willing to buy as the initial signals may have indicated. This means that the team may have to go back to Explore and examine whether they got their understanding of customer needs wrong.

There are also times when a product is scaling well in the markets; but a change in the environment scuppers growth. This may be a time to go back to Validate and rethink the business model. Products in Sustain can get a boost by the discovery of a new market

---

[10] Hohmann, L. (2006). *Innovation games: creating breakthrough products through collaborative play.* New York: Pearson Education.

and move back to Grow. These are all possibilities within the Lean PLC. What we bring to the table are a set of tools that allow companies to understand where their product is on their lifecycle journey and make informed decisions about what to do next.

1.9 The lean mindset

We hope you enjoy this playbook and find it useful in your day-to-day work. But before we begin our journey with you, we need to get in the right mindset. So here are some basics tips and guidelines.

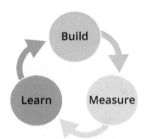

At the heart of the Lean PLC is the build–measure–learn loop.[11] This will help you identify your assumptions and test them on an ongoing basis. Across the Lean PLC, iteration and continuous improvement are key principles.

Question every assumption. No business plan survives first contact with customers. Ship experiments fast, small and soon! Test assumptions and business models before entering new markets, even with an already growing product.

Get out of the office! There are no facts in the office. Go out to the real world, meet and talk with customers. Keep engaging with the market throughout all stages of the Lean PLC. Build–measure–learn!

---

[11] Reis, E. (2011). *The lean startup.* New York: Crown Business.

Push hard as long as you can but PIVOT or STOP if you need to. It's about doing the right things at the right time. We want to be building the right product the right way; not building the wrong product the 'right' way!

You need to create cross-functional/multidisciplinary teams in which every member's contribution is valued. Keep teams as small as you can during the early stages of the Lean PLC and bring in more people as you need them.

Make decisions as a team where possible. This includes both product development decisions and investment decisions. All decision making should be based on data and learning. As your product and team grows, you have to maintain the discipline of testing and iterating.

Fail fast and learn fast. The Lean PLC is not a linear process, it's a fluid one. At any stage of the PLC it is possible to go back to earlier stages. Make these decisions quickly using evidence or validated learning.

Figure 1.5

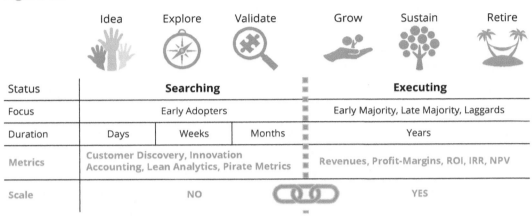

Status	**Searching**				**Executing**	
Focus	Early Adopters				Early Majority, Late Majority, Laggards	
Duration	Days	Weeks	Months		Years	
Metrics	Customer Discovery, Innovation Accounting, Lean Analytics, Pirate Metrics				Revenues, Profit-Margins, ROI, IRR, NPV	
Scale	NO				YES	

 ## An interview with:
Sebastian Cadell, Partner at Nosco

Nosco is a software and consulting company. Since 2006 they have helped some of the most forward-thinking companies bring their employees' ideas to life. This includes companies such as Danfoss, Volvo, Evonik, Heineken and Allianz. In addition to the Nosco platform, a social platform for innovation, Nosco helps clients design and run corporate innovation competitions, develop corporate entrepreneurship, build innovation communities, and leverage the possibilities of open innovation. Nosco has been cited as a leader in innovation management by Forrester (The Forrester Wave™, 2016).

In your experience, what are some of the top challenges large companies face when it comes to generating new ideas for products or services?
In most companies, only 'the select few' (the SWAT team) know about the process and really get to contribute to generating ideas. Some companies also have a stage gate approach that is too rigid and places too much emphasis on analysis and business planning in the early stages, instead of exploration and experimentation. There is often

insufficient involvement of observations, perspectives and ideas from across the organisation; and insufficient involvement of executive management in terms of decision making, commitment, risk taking, resource allocation and allowing exemptions from standard operating procedures.

As a company, what best practices do you recommend companies put in place to help their teams generate great ideas?

There should be a clear focus of the innovation effort driven by a specific and important business challenge. People should not generate ideas in a vacuum. There should be involvement of people across the organisation: the front-end and the back-end, the veterans and the newcomers, the experts and the generalists, the top and the bottom. This helps the company to benefit from multiple perspectives. Teams should not just be made up of the 'usual suspects'. Instead, there should be a process that allows for identifying relevant people in the organisation based on their ideas as well as their contributions to ideas.

Companies also need a process that allows for exploring, refining, testing and learning before passing judgement on the ideas. First ideas will often be poorly articulated and almost always be flawed. But the answer is not to have your smartest people analyse them and shoot them down. The answer is to find inexpensive ways to test the most promising ideas and learn.

Finally, there should be a process that brings ideas forward to executive management to ensure their commitment, and which motivates people to share their best ideas by letting them be part of the realisation of the idea within the company.

Without giving away too much detail, can you share a story of a company that you have worked with that ran a great idea campaign? What did they do well?

The good companies we have worked with do a lot of what I spoke about above. But they also institutionalised the process so that it became a recurring thing that could be developed and improved and could deliver better results through increased awareness, understanding and skills in the organisation. They also extended the process from innovation challenge, to implementation challenge and value realisation challenge in order to keep up the momentum in the subsequent stages of innovation.

Finally, what do you think companies should do once they have great ideas in order to actually bring them to life?

Companies need to adopt more iterative and agile approaches to bring ideas to life. They need to clearly distinguish between the process for core product execution and the process for new product development. Companies also need to provide the right structures and resources to increase the chances of success; and to keep executive management closely involved to provide air-cover and strategic support.

2

Idea

2.1 Welcome to Idea

Welcome to the left side of the Lean PLC. We begin by using our creativity to come up with great ideas. Creativity is an important part of innovation. Without ideas for products and services, companies would have nothing to put through a product lifecycle. There are many activities that can be carried out to generate ideas. Companies can choose to fund or not fund these activities. If the choice is made to fund ideation, the investment need not be large. It must be just enough to allow innovators to have the resources they need for ideation. If the choice is made not to fund ideation, then the company should have an open policy through which they source ideas from everyone. The Idea stage can be split into the following phases:

1. *Generating ideas*: This phase focuses on people coming up with as many great ideas as possible. In his seminal book *Originals,* Adam Grant shows how the best way to come up with really creative ideas is to generate a lot of ideas.[12] There are several activities that can be used to generate ideas. The ultimate goal is to prioritise these ideas and then select a few of them to work on.

2. *Capturing assumptions*: Once we have chosen the ideas to work on, we then need to capture our assumptions about the ideas. In these early stages, we will spend time thinking about who the customers are, what needs they have and how we might solve these. Our job is to capture these assumptions, identify the riskiest ones and make plans to test them during the Explore stage.

3. *Preparing for explore:* After ideation is finished, our focus will turn to getting out of the building to test whether our assumptions about customer needs and jobs to be done are correct. So at the end of the Idea stage in the PLC, we should be clear about the customer hypotheses we have and how we plan to test them. This will allow us to set a budget so we can request investment.

---

[12] Grant, A. (2017). *Originals: How non-conformists move the world.* New York: Penguin

2.2 Ideas come from anywhere

There are no limits to the sources of ideas. Ideas can come from anywhere or anyone. The best way for a company to be creative is to empower its employees have loads of ideas. Products ideas can be based on an invention or technology we already have. This is the classic solution looking for a problem. This is fine as long as we eventually do the work to make sure we are making something people want. Product ideas can also come from the market. For example, our customers can inspire ideas by the communicating the problems they are facing day to day.

Ideas can be based on:

- ☐ Our own experiences with our products and insights into gaps in the market.
- ☐ Sales teams experiences with customer requests, insights and feedback.
- ☐ Exploring trends in customer needs, competitors, markets, technology and socio-economics.
- ☐ A brainstorming session with colleagues or our companies research and development activities.

The sections that follow will describe how teams can generate ideas to potentially work on. Please note that you don't have to do everything we detail over the next few pages. You can choose the activities that are relevant to your organisation. What matters is the ability to generate and capture great ideas to take into the Explore stage.

2.3 Cross-functional teams

In most organisations, people work in silos. They are organised into specialist departments such as marketing, sales, finance, technology and HR. What this means is that people spend most of their time interacting with people who share the same type of expertise they have. In such situations, it is difficult for creative ideas to emerge. When faced with problems, the marketing team will come up with mostly marketing type solutions and the sales people will come up with sales solutions, and so on.

Sparks of creativity are generated when people work in cross-functional teams. Cross-functional collaboration exposes people to different worldviews. This gives people a chance to view the problem differently, and therefore come up with unique solutions they may have never thought of while working in their silos. Creative idea generation requires the mixture of multiple perspectives to generate unique insights and 'aha!' moments.

So every company must do whatever it takes to break down silos and promote interdisciplinary collaboration. This can be done by having people working in teams, not with their colleagues from the same department, but with colleagues from other departments. Networking events can also be organised, during which teams can learn what their colleagues are working on or struggling with. When we are planning ideation sessions or preparing to work on new ideas, interdisciplinary teams must be created.

The power of cross-functional teams goes beyond enhancing creativity. It also enhances collaboration by minimising handoffs. As they work together as team, the insights they gain are shared in real time which improves efficiency. So before we begin ideation, we need to make sure we put together a cross-functional team.

2.4 Ideation workshops

Ideas can come to people at any time. These lightbulbs moments can happen to us while taking a walk or cycling to work. If such insights strike us we should capture them and share with our colleagues at work. An ideation session is also a good way to generate ideas together as a team. These sessions can be organised regularly within any organisation. The point is to ensure that we have the right resources, information and people in place for each session.

- **Work in diverse teams –** You need a diverse team, that can bring diverse perspectives to the room. Ideation is not just for the so-called 'creative types'. Contributions by other colleagues such as finance can actually spark interesting ideas.

- **Begin with a goal in mind –** Ideation sessions must have a goal. 'Let's sit around and come up with cool stuff' is not a goal. Specific themes connected to your company's strategic goals, operational challenges or customer needs can guide the session.

- ☐ **Take time for discovery –** Once the goal for the session has been set and a diverse team has been assembled, we need allow the team time to research and gather insights. Some members of the team can talk to customers or observe them in their environment, others can research the competition, some can look at technology trends, while others look at financial and market data. We should allow the team to do whatever it needs to do to prepare for the session.

- ☐ **Have the right location and tools –** When the teams have completed their preparation, we are ready to have the session. Ideation works best with co-located teams working in the same room. Make sure that you use a room that allows for interaction and visual thinking. Teams may need walls with Ideapaint or a whiteboard, loads of post-it notes, notepads and sharpies.

With preparations complete, you are ready to run the workshop. During the actual ideation session, you can apply the following basic brainstorming rules:

1. Make sure you have a facilitator to help guide the session. The facilitator must understand ideation and have the capacity to draw participation from individuals that would normally be quiet during team meetings.

2. Before ideation begins, make explicit the goals of the session, so that people keep focused. The facilitator can gently nudge people if they go off-track.

3. To get people engaged and ready, use fun warm-up exercises. For example, you could have the team come up with as many uses for a bin-bag as they can think of.

4. When ideation starts, it is better for individuals to first work alone, before they engage with the team. So, restate the goal for the session, and then have individuals come up with ideas on their own before collaborating with others.

5. Encourage teams to capture their ideas visually. Napkin sketches of ideas on large post-it notes are a great way to stimulate conversation.

6. When individuals start sharing their ideas, the facilitator should enforce brainstorming rules, such as deferring judgement and having only one conversation at time.

7. All ideas that a generated should put on a whiteboard or a wall for the team to review and prioritise. You can use dot-voting to prioritise ideas. When voting, individuals should have a limited number of votes to use on ideas (e.g. three votes).

8. After the team do their voting, it might be wise to bring in important leaders in the company who may have a say in ideas going forward to get their input. These leaders can then cast the deciding votes.

At the end of the session, decisions must be made about which ideas are going forward for further review, analysis and preparation for Explore.

2.5 An open idea campaign

Another great way to gather ideas within a company is to run an idea campaign. This campaign is typically an open call for every employee to submit any ideas they have for the company to work on. Idea campaigns can be great sources for ideas, but they have to be managed well.

☐ **Open to everyone –** The idea campaign should be open to everyone within the company, not just product people. Gather ideas from as many people as possible; allow people to submit more than one idea.

☐ **Easy to use –** Create an easy to use platform that allows employees to easily submit, vote and comment on ideas.

☐ **Clear goals –** The campaign must have clear goals that are communicated to all employees.

☐ **Clear outcomes –** Make clear what the outcome of the campaign will be. If the winner is going to get investment to work on their idea, make this very clear to employees.

☐ **Make decisions –** Review the ideas submitted and make decisions on the winning idea based on predetermined campaign goals and criteria.

☐ **Keep promises –** Be accountable and keep promises made to the winners. This is critical to keeping employees motivated for future campaigns.

☐ **Support community –** Maintain and support the community after the campaign is over by providing resources, tips and updates.

Open idea campaigns should not be simply used as a fun exercise to get 'employee engagement'. After a few such campaigns, you may find that employees become sceptical. Instead, companies should use open idea campaigns as an authentic way to gather ideas that will eventually be turned into products. This is why these campaigns should be informed by strategic goals that are made explicit to all employees. At the end of the idea campaign, there should be resources to at least take the idea into the Explore stage of the lifecycle.

2.6 Techniques to spark creativity

During ideation sessions and open idea competitions people can get stuck for ideas. Coming up with ideas is difficult to do in the best of circumstances. It is much harder to do in the day-to-day grind of our daily work. As such, in addition to cross-functional collaboration and providing teams with strategic guide, the following techniques can also be used to spark creativity.

1. **Customer insights**

   An excursion to the see customers in their contexts can be a rich source of insights. Teams can also speak to sales to find out 'What customers have been telling us.' During ideation sessions, we can have teams capture the insights on post-it notes and place them on a wall. They can then discuss the insights and come up with as many ideas as they can.

2. **Technology**

   Sometimes companies have technology that has been developed within your organisation but is yet to be commercialised. This can be used as the basis for an idea campaign or an ideation session. Teams can also research emerging technology trends and use these as basis for ideation.

3. **The environment**

   Beyond technology, other aspects of a company's business environment can also be a rich source of insights. For example, teams can then spend time researching any economic, social, political, legal and competitor trends that may impact the business. When this task is complete, the key question becomes whether our current business model is adaptive to its changing environment. The teams can then imagine how they might redesign the business model to be more adaptive.

4. **What ifs**

During ideation workshops, we can also use a series of what-if questions to trigger ideation. What if we had to give our products away for free? What if we lost our largest client? What if we could no longer use the Internet? Come up with as many 'what ifs' as possible; they don't have to be realistic. The point of the exercise is to have the teams ideate about what they would do to make the business work if the 'what-if' scenario came true.

5. **The ideas box**

For open idea campaigns, it might be helpful to provide teams with resources they can use for ideation. An example is Adobe's Kickbox,[13] which is a red box that is available to all employees to pick up and use. Inside the box you will find a set of cards, tools and notebooks that provide a step-by-step guide of what you need to do to generate, prototype and test an idea. The power of the Kickbox lies in the fact that it is available to anyone in Abode; with a prepaid credit card with $1000 on it!

2.7 Prioritising ideas

The point of all the activities, methods and tools we have described so far is to create an environment that is buzzing with collaboration and insights. In such an environment, people will generate a lot of ideas. However, it is impossible to work on every idea given the limited resources that every organisation will have. Some prioritisation of ideas will be necessary. There are several filters that can be used to choose ideas, including the following:

☐ *Is the idea aligned to the company's strategic goals?* Ideas that lack alignment often end up as orphans that no one is willing to take to scale.

---

[13] Wilson, M. (2015). *Adobe's kickbox: the kit to launch your next big idea.* Fast Company: Available at https://www. fastcodesign.com/3042128/adobes-kickbox-the-kit-tolaunch-your-next-big-idea

☐ *Does the idea help us meet our portfolio goals?* Every company should have a balanced portfolio with products covering core, adjacent and transformational innovation.

☐ *Does the idea help the company manage and adapt to emerging trends in the business environment?* There may be key changes in the world that are impacting our business and we can choose ideas that help make the company more adaptive.

☐ *How large is the potential market for the idea?* This is the least important criterion for early-stage ideas. Small nascent markets sometimes grow to become important over time. However, after using the criteria above, market size can be used to make a final choice from remaining ideas.

☐ *Which ideas got the most the votes?* During ideation sessions or idea campaigns it is possible to have employees voting for the ideas they like. For example, earlier we spoke about dot-voting during an ideation session. These votes should be considered when making prioritisation decisions.

Whatever method we end up choosing for prioritisation, the end goal is to have one or two ideas that teams can focus and work on. Once an idea is chosen, we can turn our attention to examining the idea for assumptions.

2.8 Capture plan A

Now that our team has chosen a great idea to work on, it's time to capture our assumptions. We are sorry to break it to you. Your idea is not as great as you think it is. As innovators we often think that our idea is the best thing since sliced bread; and we are often assuming that customers will feel the same way. So the expectation after choosing an idea is that you are going to write a beautiful business plan and get vast amounts of money to work on your idea.

But no business plan ever survives first contact with customers, and it is them we making the products for. If they don't like our product then it is dead in the water. So as much as you would like to move straight to building your product, we encourage caution. Rather than write a business plan and get approval for a large investment, we would first like you to capture your plan A on a canvas. At this early stage, we are mostly interested in the assumptions that you are making about customers, their needs and the potential solution you will be creating for them.

Our goal in these early days is to make sure we are making something people want. This is the number one goal in innovation. If we successfully achieve this problem-solution fit, we can then do the work of figuring out how to deliver value to customers with a profitable business model. However, if we are making stuff no-one wants, then it does not really matter how clever our business model is.

As first step, we recommend the using a tool such as the Value Proposition Canvas (VPC).[14] The reason we like this tool in these early stages is that if focuses on customers. It allows teams to capture their initial assumptions about customers' needs and jobs to be done

2.9 Using Value Proposition Canvas

The Value Proposition Canvas (VPC) was developed by Alex Osterwalder and colleagues. It is a deep dive into the 'customer segment' and 'value proposition' sections of their Business Model Canvas. It is useful as an early tool, when you are not yet clear on the other details of your model. You can visualise your assumptions about customers and their needs early in your process. We will provide tips on how to use the canvas below.

Below are tips on how to use the Value Proposition Canvas:

- ☐ Use sticky notes, do not write directly on the canvas.
- ☐ Use words and images to illustrate your assumptions.
- ☐ Put one assumption per sticky note – don't make lists.
- ☐ Do it together as a team and ensure that every team member contributes.
- ☐ Sketch alternative jobs to be done so you can think more widely about your value proposition.

---

[14] Osterwalder, A., Pigneur, Y., Bernarda, G. and Smith, A. (2015). *Value proposition design*. New York: John Wiley.

Figure 2.1

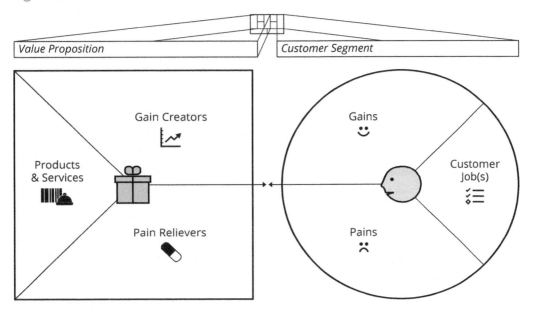

- Focus on quantity over quality, you can review the assumption later.

- Map separate canvases if you have more than one customer segment.

- Do not spend too much time mapping the canvas. One hour is more than enough. Remember we are just capturing assumptions that we will test later.

When completing the canvas, it is better to do this as a co-located team. In these early stages of Idea, a team need not be more than three people (e.g. a designer, a product owner, a researcher). You can create an A1 or A0 size version of the canvas and place it on a wall. It is also important to make the session fun and engaging.

2.10 Map the circle – map the square

When most people are asked to describe their value proposition, the first thing out of their mouth is a description of their product and what it does. There can be no 'value proposition' unless your product creates value for someone. Customers don't care about your product (unless they are your mom), they only care about how the product meets their needs. This is why when working on our value proposition we first map the circle (i.e. customer profile).

- *Customer jobs to be done*: This part of the map focuses on the things that customers are trying to achieve in their lives. In other words, what would make customers reach out into the world and grab your product. Jobs to be done can be functional, such as tasks that customers want to complete. They can also be social, personal or emotional, such as achieving social status or feeling accomplished. We need to map as many jobs as we can think off.

- *Pains*: There will be pains and difficulties that connected to the jobs that customers are trying to get done. These can be obstacles, problems or undesired outcomes. For example, young children might find it difficult to use a mouse owing to their limited motor skills.

- *Gains*: If customers accomplish their jobs, what gains would they have created in the lives? This focuses on the outcomes that accrue to customers if they successfully complete their jobs. The gains we map can be expected gains, required gains or desired gains. We may even have more impact if our product creates unexpected gains.

Now that we have mapped our customer profile, we should next turn our attention to completing the value map. At this point, we should have very loose ideas about what solution we might be thinking of creating:

- *Products and services*: On this section, map a simple list the products or services you think you might offer customer to help them get their jobs done. These can be physical goods, services or digital products.

- *Pain relievers*: Pain relievers are connected to the pains you mapped out in the customer profile. In this section of the canvas, you should map out how your proposed solution may solve the pains.

- *Gain creators*: Similarly, gain creators are connected to the gains you mapped out in the customer profile. In this section of the canvas, you should map out how your proposed solution may create the gains that customers want.

Please note that it is not sufficient to simply mirror the pains or gains and then say that your product will solve for them. You have to map out how this will be done. For example, if the pain that students face is how to get to school because of long travel distances, you cannot simply say that your product will solve the distance problem. You have to say how; for example, we will provide regular school buses.

2.11 Identifying risky assumptions

After thinking through your chosen idea, you are now in a position to identify your risky assumptions. On your value proposition canvas, you will have mapped some jobs, pains or gains that have already been validated in other contexts. However, on your canvas will also be a lot of untested assumptions. For example, you may be attributing pains to customers that you may not have evidence for. If you are wrong, then this will increase the likelihood that you will build the wrong product.

As such, the team must step back after mapping the canvas and review each post-it note. Only one questions should be asked: Do we have evidence for this or is it an assumption? When in doubt, it is an assumption. Using this process, the team can:

☐ Identify all your untested assumptions. You have to be honest. There is no need to fool yourself and declare something as known, when it is in an unknown. Put a large red X on all sticky notes with untested assumptions.

Figure 2.2

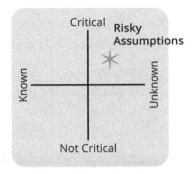

35

Q Useful tips

- Work as a team and interrogate each assumption fully.

- Look at your competitors and see if any have validated some of your assumptions. Look outside your industry for other evidence of validation.

- Identify your most critical untested assumptions and make plans to get out of the building and test those first!

- On a scale of 1–10 (1 = *Not Critical* and 10 = *Critical*), rate each untested assumption in terms of how critical it is to the success of your idea (i.e. if it turns out you are wrong on this assumption, you would have to stop the project). The most critical assumptions are your riskiest assumptions. These are what we will test first during Explore.

2.12 The market

After mapping your Value Proposition Canvas, there are two key questions you need to be able to answer as you prepare to go into Explore:

- Do we have any products in our company that are already serving the customer needs you identified? If yes, then why does your company need to work on your proposed solution? What is the extra value it will provide?

- What is the nature of the competitive landscape? Are there competitor products serving the customer needs we have identified? If yes, then why should your company enter this space? What is the unique value your solution will bring to a market with competitors in it?

These two questions require you to do some research in order to complete. During Idea, extensive research is not expected. We recommend that your team does some lightweight research just to get a sense of the landscape. During Explore, you can conduct more extensive research:

- Talk to colleagues within your company to find out if you already have similar products in the market.
- Look at key business publications for key trends and macroeconomic factors.
- Study competitors within your industry offering similar value propositions.
- Look outside your industry for competitor products that make good alternatives.
- Identify key gaps in knowledge that you want to examine further during Explore.

At the end of this process, you will have a sense of any outstanding questions that require further research during Explore. You can now turn your attention to making plans for testing your ideas.

2.13 Brainstorm tests

At this point, your team will have come up with several ideas; prioritised those ideas and chosen one to work on. Your team will also have reviewed your chosen idea using the Value Proposition Canvas for assumptions; identified and ranked the riskiest assumptions. Your team will also have spent time analysing their knowledge of the market and identified gaps that need to be researched. This means that the team has a list of untested assumptions about customer needs, potential solutions and the market. During Explore, the team will be getting out of the building to test these assumptions. In preparation for Explore, your team needs to spend time brainstorming ideas of how to test your risky assumptions.

We will provide details of the different types of tests and experiments you can run in the next chapter. At this moment, it is sufficient to say that there are four main types of test a team can run during explore:

- *Desk research*: This is useful when testing assumptions about the market. You can read business reports and published papers on market trends.

- *Customer observation*: Your team can spend time observing customers in the real world. You can also be a *participating observer* by working with customers and helping them achieve their goals.

- *Customer interviews*: Some of your assumptions can be tested by talking customers. The discipline is to make sure that your questions are unbiased.

- *Customer action*: Finally, you can run experiments during which you can ask customers to perform some action (e.g. download an application).

As a team, you need to spend time reviewing each assumption and coming up with ideas of how to test it. If an assumption requires a large study to test it, break it down into smaller tests. When you are done brainstorming, you need to prioritise the set of tests you will run during Explore. When you have some ideas what you want to do, you can then think about what resources you will need for the Explore.

2.14 Preparing for Explore

Within the Lean PLC, products are moved from stage to stage at the decision gates. This is particularly the case during the search phases of the Lean PLC. There are key criteria that each team has to meet in order to pass each stage gate. If you have done the activities we described above, your team is pretty much ready to ask for funding from the Product Council to move to the Explore stage.

A key rule for the Lean PLC is that teams should not be asking for large amounts of money to build their products at this stage. Instead, there should be asking for a minimal investment to test their key assumptions about customer jobs, pains and gains during the Explore stage. Our investment request may include:

☐ Brief details about the idea we plan to work on.

☐ How the idea supports the strategic goals of our company.

☐ Our assumptions about customer segments and their needs.

☐ Our assumptions about the market (potential size and competitors).

☐ The assumptions we plan to test during Explore, and brief details of the tests we plan to run.

☐ The resources and funding you think you will need to complete all the tests you want to run during the Explore stage.

This is all the information you need for your submission, not a 40-page business case. The product council or investment board will then review the application placing particular weight on three key things: alignment of the idea to company strategy, the clarity of the proposed assumptions and the planned tests for Explore.

SUBMISSION TEMPLATE
Idea to Explore SUBMISSION

Idea ownership

Investment board	
Business sponsor	
Product owner	

Idea overview

Product name	
Idea description	
Strategic fit	

Target customers and jobs to be done

Describe your potential customer segments and their needs or jobs to be done.

Customer segments	Jobs to be done

Assumptions to test during Explore

Provide an overview of the assumptions about customer needs you plan to test during Explore, how you will test them and your success criteria.

We believe that:	
To verify that we will:	
And we will measure:	

| **We will know we are right if:**[15] | |

Market assumptions

We are currently assuming that our idea serves this market opportunity.

| |
| |

Resources and funding requested

To complete the Explore stage, we are asking for (e.g. **dollars, time, people, etc.**):

| |
| |

We need the resources to do the following during Explore:

| |
| |

[15] Osterwalder, A., Pigneur, Y., Bernarda, G. and Smith, A. (2015). *Value proposition design*. New York: John Wiley.

An interview with:

Nikitas Magel, Senior Content Manager at Brightidea

Brightidea, the industry leader in innovation management solutions, has a mission to transform how the world innovates. To that end, it builds tools to facilitate and streamline the ideation process, developing the right mix of specialized software and curated services to drive success at every stage of the innovation journey. Through its unique and highly versatile apps, the Brightidea Innovation Cloud enables corporate innovation teams to set up, launch, and compare initiatives, and centrally track their impact. Innovation team leaders can then measure and maximize the financial return of their efforts, empowering them to justify their programs, innovate further, and think bigger.

In your experience, what are some of the top challenges large companies face when it comes to generating new ideas for products or services?
Probably the biggest challenge for any large company is achieving a creative monopoly. Doing so requires focusing the business on developing something entirely new rather than copying something that already works. But large companies often focus instead on incrementally improving an existing product or service. That approach only creates more competition, resulting in several companies vying for a share of the same market. With a creative monopoly, driven by technological innovation, a company can differentiate itself in the market to such a degree that it preempts any competition and generates a continuous flow of profit.

One important approach to creating a monopoly is to ask what's at the core of the customer's need and if there's something that can go deeper in satisfying it. That requires empathy. Which means making an effort to understand people: how they do things and why, what their physical and emotional needs are, how they think about the world, what is meaningful to them, and what makes their lives easier and more enjoyable. For large companies used to making incremental improvements on established techniques, this empathic approach is especially challenging – because it means questioning underlying assumptions and being open to eschewing them, which many are reluctant to do.

As a company, what best practices do you recommend companies put in place to help their teams generate great ideas?
For any company that's serious about innovation, we believe it's critical to develop a culture of innovation. That's because ideation (and the creativity behind it) can't be

encapsulated or isolated; it's far more effective when it permeates the organisation. It has to be systemic. While a company might have a dedicated team to manage innovation initiatives, the real engine of innovation is the organisation at large, whose creative fuel has to circulate throughout it in order for that engine to hum. Implementing a culture of innovation means strongly encouraging ideation among employees at every level of the organisation, and providing the framework, tools, time and permission for them to do so.

Without giving away too much detail, can you share a story of a company that you have worked with that ran a great ideas campaign? What did they do well?

One company that manufactures automotive safety equipment recently ran into a problem with seat belt mechanisms. In the process of solving it, they set out to understand the problem very deeply, then came up with a set of constraints by which a successful solution would work, and finally pulled in the right people to come up with solutions. That last part is key, because it wasn't just the experts in seatbelts who were involved. Bringing in people from other, even unlikely, areas of expertise presented perspectives that ultimately helped to solve the problem, which the seatbelt experts alone couldn't. This showed that for ideation to be successful, it's important to widen and diversify the source of ideas, which is where the power of crowdsourcing lies.

Finally, what do you think companies should do once they have great ideas in order to actually bring them to life?

The ideation process involves great deal of friction. There are often different perspectives that result in competing ideas and opinions on whether and how any of those ideas will work. The success of ideation depends on how early in the process that friction occurs. The later it happens, the more likely an idea will advance only to be shot down after time, energy, and resources have already been invested. In that case, a product might get far along in its development, then nixed when a key influencer or decision maker – who is consulted only towards the end merely as a 'sign off' – shines light on an underlying problem. But when creativity is activated and different perspectives are brought in at the beginning of the ideation process, that friction is smoothed out gradually as an idea develops, allowing it pick up momentum, gain viability, and be successfully implemented.

3

Explore

3.1 Welcome to Explore

Welcome to Explore. You've pitched your idea and it has been approved for further exploration. The main goal of Explore is to obtain evidence that your customers have a real need or problem. You will now get out of the office and speak to customers to test your assumptions. At this stage you do not yet need a large team. All you need is a small team with sufficient knowledge around the market and customers (e.g. a product manager, a UX (user experience) expert or designer, with some support from sales). The Explore stage can be split into the following phases:

1. *Validate problem:* This phase focuses on customer needs and jobs to be done. Do customers have the need? How are customers currently solving their problems? Are the problems currently serious enough for them to pay for a solution?

2. *Value proposition:* After validating customer needs, we can start testing our ideas for a solution. We can begin by creating experiments that present potential solutions to customers (e.g. landing pages). At this stage, we are not creating our product yet, we are simply testing our ideas for solutions and value propositions.

3. *Market opportunity:* When we validate customer needs, we will know the type of customer and market we want to pursue. As such, during Explore we will do more detailed research on the size of the total available market and whether that market is financially viable.

4. *Prepare for validate:* At the end of Explore, you will have confirmed customer needs and also tested some ideas for a solution. In Validate, you will begin working out how you can create a solution to sell and test the rest of your business model. So you need to define the budget for building a minimum solution and its various iterations, market it to customers and support them through their experience.

3.2 Running experiments

A key part of lean thinking is the use of evidence to make decisions. For many, innovation and product development are about the pursuit of a vision. Given the failure rates of startups and new product launches, the idea of vision alone leading to success seems more of an illusion than anything else. There is nothing wrong with chasing big audacious goals. However, our passion for our ideas must be tempered with a keen interest in reality (i.e. testing our assumptions).

To test our assumptions, we need to adopt the scientific method and apply it to innovation. The idea of science conjures up images of men and women in white coats working in a lab. However, science is merely a set of tools and methods. When we test our ideas, we are not merely collecting evidence and data. We are trying to use the evidence as a basis of learning and decision making.

Science is a hypothesis driven methodology. We cannot use the data we collect to learn and make decisions, unless we first make explicit our expectations. As such, we must turn the assumptions we identified during Idea into testable hypotheses. The best way to create hypotheses is to set minimum success criteria for our experiments. These criteria benchmark what would have to happen for us to reach the conclusion that the evidence supports our assumptions. With clear hypotheses in place, any data we collect can then be benchmarked against our criteria. This makes it easy for us to learn, make decisions and track progress.

It is also important to note that minimum success criteria cannot be developed in a vacuum. Such criteria are connected to a specific test or experiment. So, the first thing we have to do for each assumption we have is brainstorm some experiments. After we have chosen a specific experiment to run, then we can set the minimum fail criteria for that experiment. Only after this process are we now ready to get out of the building and start testing our assumptions.

3.3 Tracking progress and managing our experiments

When we run our experiments, we need a tool to track the progress of the work we are doing. Tracking progress happens at the level of a single experiment we are running, and also across experiments to see how well we are doing at reducing the number of assumptions in our value proposition or business model. For the Lean PLC, we have adopted the Experiment Canvas.[16]

1. **Assumptions:** Which specific assumptions are you testing? What are you trying to learn? How important is this assumption to the success of your product or service?

---

[16] Viki, T., Toma, D. and Gons, E. (2017). *The corporate startup.* Deventer: Vakmedianet.

2. **Experiments**: What experiment are you running? How are you going to conduct the experiment? What exactly are you going to do?

3. **Cohort:** Who are the people, businesses or customer segments that you are going to be running the experiment on? It is also important to think about where you will find the participants for your experiment and also whether you have any recruitment criteria.

4. **Success criteria:** What are your minimum success criteria? What would have to happen in your experiment for you to decide that your assumptions are supported by the data? How will you decide to pivot, persevere or stop?

5. **Timebox:** When running experiments speed is critical. As such, we need to set an upper limit for the time we want our experiment to run.

After we have mapped out the experiment, we are now ready to get out of the building and test our assumptions. After we have finished running the experiment, we then have to analyse the data we have collected and make decisions.

6. **Results:** At this point, we analyse and tally what we found? For example, how many people choose one option versus another?

7. **Learnings**: Now that we have tallied our data, what did we learn? How does the data match up to our minimum success criteria? What unexpected things did we learn when we were out of the building?

8. **Decision:** We are now ready to make a decision. What do you do now that the experiment is a success/failure? Do we want to pivot, persevere or stop? We also have the option to run other experiments using a different method or cohort in order to further test our assumption.

🔍 Tips

The key is to work as a team when designing experiments and setting minimum success criteria. The team must also come together to review the data, learn lessons and make decisions.

☐ Always start with your riskiest assumption. Don't try to avoid it!

☐ Use tools to track every experiment and its results. This way you can always track the decisions you are making as you go.

☐ If your first experiment doesn't work, you can redesign it and try again.

☐ It is a judgement call, but sometimes you have to know when to stop. For example, when you begin to hear the same things from customers, stop!

Figure 3.1

Experiment Canvas

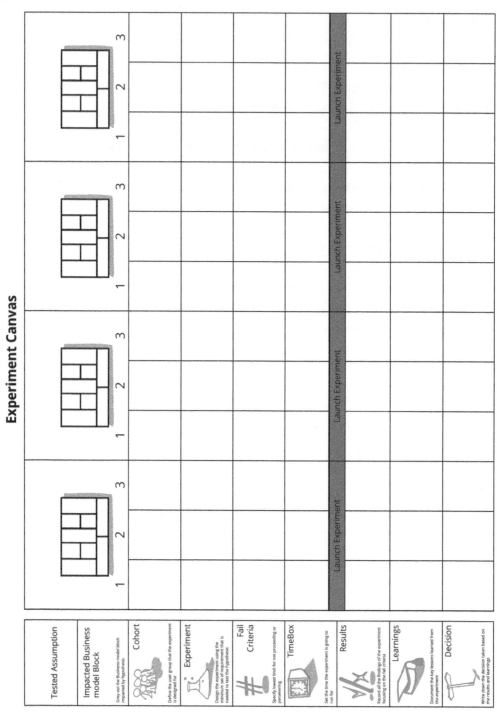

Tested Assumption			
Impacted Business model Block			
Cohort			
Experiment			
Fail Criteria			
TimeBox			
Results			
Learnings			
Decision			

3.4 Earlyvangelists[17]

An important question for running experiments concerns who should be recruited as participants. There are also related questions about sample size and how we know when to stop running an experiment. First, experiments during the Explore stage should target early adopters or earlyvangelists. Working with earlyvangelists[18] has the potential to provide the greatest amount of learning because:

- they have the problem or need we are trying to solve
- they are aware that they have the problem or need
- they have been actively looking for a solution
- They have tried to put together a solution
- they have a budget to pay for a solution.

These characteristics make earlyvangelists, if we can find them, the best people to test our assumptions early in our process. Earlyvangelists are so keen for solution that they would be happy to give us feedback on early versions of our product. It is also important to note that if we find a customer segment in which any one of these five characteristics is missing, then it will be difficult for us to test customer needs and validate early versions of our solution. For example, if we find that customers have a

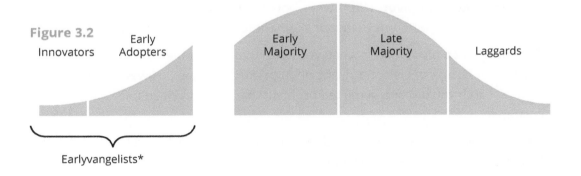

Figure 3.2

Innovators | Early Adopters | Early Majority | Late Majority | Laggards

Earlyvangelists*

[17] Moore, G.A. (1999). *Crossing the chasm*. New York: Harper Business.

[18] Blank, S. and Dorf, B. (2012). *The startup owner's manual*. California: K & S Ranch.

need but they are not aware of it, we now have a marketing job on our hands, which is the opposite of the discovery we want to do during Explore.

With that in mind, we can turn to the question of how large our sample of participants should be. We do not have any specific guidance on this. It really depends on the type of research you are doing (e.g. B2C vs B2B). Our sense is that experiments with earlyvangelists should provide strong signals. As such, a small sample of 10–15 participants should be sufficient. If you find that you are getting mixed signals, you may need to define more stricter criteria for selecting your earlyvangelists. In contrast, if you are getting a very strong and consistent signal from customers, you can stop before you reach your target sample size.

3.5 Pivot, persevere, stop

We run experiments to be able to make decisions about what to do next. We have the choice to persevere with our idea if the evidence supports our assumptions. However, sometime the data suggest that our assumptions are wrong. In that case, we don't want to just give up on our idea. We also don't want to keep going by simply ignoring the evidence. We can make the decision to pivot. This is when we keep one foot in what we have learned, and we make some changes to our assumptions before we test them again. In *The Lean Startup,* Eric Ries[19] described ten types of pivots that teams can make if they find that their assumptions are not supported by the evidence.

1. **Customer segment pivot:** This is when you find out that your product solves a real problem, but for a different customer segment to what you initially assumed.

2. **Customer need pivot:** This is when you learn that the problem you were trying to solve for customers is not a strong enough need for them. However, during your research you discover a related problem that is truly a need for your customers.

3. **Zoom in pivot:** This is when, something you were thinking of as just a feature in a bigger product, becomes the whole product.

4. **Zoom out pivot:** This is when, what you thought of as the whole product, becomes a feature in a bigger product.

---

[19] Ries, E. (2011). *The Lean Startup.* New York: Crown Books.

Figure 3.3

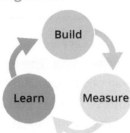

5. **Platform pivot:** This refers to a situation where you change our idea from just being an application to being a platform, or vice versa.

6. **Business architecture pivot:** This is when you change your idea from being a long margin, high volume product serving consumers (B2C) to a high margin, low volume product serving businesses (B2B), or vice versa.

7. **Value chapter pivot:** This is when you change the revenue model for your product. For example, you may have been thinking about having customers subscribe to your service, but after talking to customers you learn that they are much happier to make a one-time purchase.

8. **Channel pivot:** This is when you change your initial ideas about the sales or distribution channels you would use to reach your customers.

9. **Technology pivot:** This is when we learn that we can deliver value to customers using a different technology to the one that we intended to use.

10. **Engine of growth pivot:** In Chapter 5, we will discuss how to grow and scale our product. In that chapter, we will present the three main engines that teams can use to grow customers numbers, revenues and profits. For now, all we need to note is that we can use evidence to change our initial growth engine to a different one.

These are some of the options that we have when we are faced with a situation where some of our assumptions are not supported by the data. We are often asked how many pivots a team should make before they stop. This is a judgement call that depends on the strength of the signal you are getting from the market and the resources you have at hand. If the resources are available you don't want to stop too soon. However, if you are getting really strong signals that your idea will not work, then don't waste your resources. Stop and move on to another project!

3.6 Desk research

Desk research, also known as secondary research, is a good starting point for testing our assumptions before we get out of building. The focus here is on information that already exists. Although the data may not be specific to our hypotheses, they are great for learning about the market and our customers before we meet them face-to-face. In this age of ubiquitous information, it is amazing what data are readily available for us to review and make informed decisions.

Desk research is particularly useful for testing our assumptions about the structure and size of the available market. It is also useful for discovering any trends in population, economics and technology that may impact our product or service. We can use desk research to get information on demographic trends, suppliers, distributors, competitors, brands and products. All of these data can give us a good sense of the market environment we are about to enter.

How to do it

Desk research can be viewed as a journey of discovery. You just never know what you are going to learn. At the minimum, we need to be clear about the assumptions we want to test and then set clear learning goals. When we are ready, there are several sources of secondary research we can use.

☐ *Market research reports*: Such reports can be obtained from market research and consulting companies such as CB Insights, Nielsen, Ipsos, Deloitte and McKinsey. They are usually great sources for information around key trends.

☐ *Government statistics*: Most countries have a national statistics office. Some, like the UK government, even publish their data online. These websites allow you to research industry-based topics such as energy, education, agriculture and the labour market.

☐ *Industry experts and bodies*: If you are thinking entering a specific industry with your idea, it may be useful to read any research published by experts in that

---

industry. There are also bodies that represent the interests of companies within a particular industry. Such bodies often conduct research and have industry-based publications that can be rich sources for learning.

☐ *Company data*: This is a good way to research the viability of particular types of business. Company data can be obtained from government run agencies such as Companies House in the UK or the US Securities and Exchange Commission. These data allow you to learn about company registrations and closures, as well as their company accounts and annual returns.

☐ *The Internet*: This is also a rich source of data. For example, you can use Google Trends, Google Keyword Planner and other Google keyword research tools, to find out how often people are searching words related to your product. Social media analysis can also be useful for tracking trends on particular topics and understanding the language that customers use around their needs and jobs to be done.

🔍 Tips

☐ Avoid analysis-paralysis. There is so much research out there that it can be overwhelming to make sense of it all.

☐ One way to stop yourself from running down every rabbit hole is to make sure that you have a learning goal. This then allows you to focus on only that research that will help you with this goal.

☐ Use desk research as starting point. By itself, secondary data can provide good signals, but it is not conclusive. Secondary research can point you in the right direction, but you still need to get out of the building and talk to people.

3.7 Customer observation

Genchi Genbutsu is a Japanese phrase that means 'actual place, actual thing'. It has been popularised as a principle of Toyota's Production System.[20] The principle is based on the notion that in order to understand something you have to go to the actual place and see the actual thing you are interested in. This is why the phrase is often translated into English as 'go and see for yourself'. As some point, we have to get up from behind our desks and see how our customers live and work in their context. Customer observation allows us to gain empathy into what our customers struggle with. Such insights can then be used to inform product development.

How to do it

When you are out observing customers, you can select from a toolbox of observation methods.

☐ *Silent observation*: This form of observation is unobtrusive. It is most powerful when people don't know you are there. Unlike chemical compounds, people tend to respond to being watched by others and may change their behaviour as a result. So as long as you are not violating privacy laws, it may be helpful to observe people silently. You can also get prior consent from individuals and organisations to do such observation.

☐ *Conversational observation*: The challenge with silent observation is that when you observe something interesting, it is difficult to find out the underlying reasons for the behaviour. This can leave us with uninformed assumptions. To get a deeper understanding of our customers, we can have conversations with them while we are observing their behaviour. A common technique in user experience research is to have customers talk through their thinking while they are doing a task. This can help with both seeing what they do, and understanding the reasoning behind it.

---

[20] Liker, J.K. (2004). *The Toyota way.* New York: McGraw Hill.

> ☐ *A day in a life shadowing*: Rather than just watch people while they complete a single task, you can ask for permission to shadow them the for a whole morning or day. This is particularly useful within work environments, where your product might be used in a variety of different situations.
>
> ☐ *Participant observation*: Observing and speaking to customers is good way to gain insights. However, to gain even deeper insights we have can participate in the tasks that our customers are trying to complete. This allows us to walk a mile in their shoes and truly understand how difficult it is to do some of the jobs they are trying to get done. So instead of just watching, try it yourself!

Rules of the game

Customer observation can be a bit like drinking from a fire hose. There is often a lot to see, hear and capture. What we conclude from what we see can also be biased by our previous expectations. In particular, innovators can have a bias towards confirming their own ideas. As such, it is important to follow a few rules of observation so that we can collect reliable data that are useful for making decisions.

☐ *Learning goal*: Always begin with a learning goal. What are you trying to learn or find out? Making this explicit at the beginning ensures that we don't get distracted by irrelevant observations.

☐ *Outline questions*: Even if you will not be speaking to customers, you should prepare a set of questions that you will try to answer from the observation. As you see and hear things that answer your questions, you can then capture the data in the appropriate places.

☐ *In context*: Observe customers in their context. This help you to understand their jobs to be done more fully. You will get to see what happens before, during and after the task. You will also gain an understanding of the constraints that exist within the context that the jobs are being performed.

☐ *The mundane*: When observing, don't seek to find the extraordinary. A lot of insights lie in things that seem mundane. So, capture what you see without selective bias, even the ordinary events. As long as it is related to task you are observing, such data can help you meet your learning goals.

☐ *The outliers:* At the same time, keep an eye out for the outliers. These may be people who are heavy users of your products or people who suffer from the problem you want to solve in an extreme way. Outliers can be helpful in allows us to determine how far we might have to go when creating our solution in terms of a minimum feature set.

☐ *Find example solutions*: It may be important to capture examples of the artefacts that customers currently use to solve the problem or meet the need they have. These could be competitor products or hacked up solutions. Such artefacts can be helpful when we start designing our solution.

☐ *In pairs*: Where possible, conduct your observational research in pairs. This is particularly important when conducting conversational observation. One person can interact with the customers, while the other captures the data.

☐ *Capture and externalise*: Capture all your observation within the moment they are occurring. Do not wait to do this later as you will have to rely on your memory. If you wait till later, you will forget key information. And if you have observed more than one customer, you may misattribute behaviour to the wrong person. Capture what you see with fidelity. Don't analyse it in the moment. You can do data analysis later.

When you have finished your observation, you can then review all the information you have captured. You can use sticky notes or flashcards to organise your data. Capture each insight from your research on a separate sticky note or flashcard. Organise the cards or sticky notes on a wall or table based on themes. Use this process to identify patterns and anomalies in your data. After analysis is complete, communicate your learning to your team.

3.8 Customer interviews

We already highlighted the limits of simply observing our customers without talking to them. While we can learn a lot from watching people, it is often difficult to understand why people are making specific choices or taking specific actions. Furthermore, it is sometimes difficult to get access to observing people in their context. There may be constraints in terms of time, availability or even legal issues. In such circumstances, customer interviews are a quick and cheap way to gain access to customers and learn about their needs.

How to do it

There are several issues to consider when interviewing customers. Our goal is to get reliable data that we can use to make decisions about possible solutions.

☐ *Choose your participants*: A key discipline to have is to make sure that we don't just talk to anyone who will talk to us. It is important to identify early adopters and speak with them. Using the five criteria for earlyvangelists, we can craft filter questions that allow us to select the right people to talk to.

☐ *Talk to experts*: Another source for useful information are experts within the area that you are exploring. Talking to experts in particularly helpful when you are thinking of solving a problem in a space in which domain or specialist knowledge is necessary.

☐ *Use scripts*: Before talking to customers, you need to be clear about what you learning goals are. On the basis of these goals you can then create an outline of the interview with questions that are aimed at testing your assumptions.

☐ *Go off track*: Even with a script, customers can often unexpected insights. Allows the interview to go off track every now and again. Capture the new insights and then return to your script.

☐ *Get connected:* At the end of every interview, ask customers whether they know other people facing similar challenges that they can introduce you to. This snowball method is a good way to find early adopters for our solution.

☐ *Capture data*: During the interview you have to capture all the data and learning that you can. One way to do this, is to ask for permission to record the interview, and then you transcribe and review the script. Another way is to do interviewing in pairs.

☐ *Review immediately*: At the end of every interview, review the notes you have taken and highlight key learnings before you move on to the next interview.

The Mom Test

Talking to customers can be a great source of insights and learnings. However, when it is not done correctly, it can also be a source of false signals and misleading information. First, people have a tendency to want to make a good impression on other people. This social need may lead them to answer questions in way that they think will please the interviewer or make them look good. Second, people may find it difficult to introspect and find the right language to communicate their feelings. Often people don't have an opinion and may feel pressured to create one because of the interview. Finally, people are terrible at predicting what they will do in the future, even if they appear to be confident when answering your questions.

These challenges in talking to customers led Rob Fitzpatrick to write a book entitled *The Mom Test*.[21] In the book, Rob argues that innovators should not ask customers questions that if they asked their own mum she could lie to them. Instead, they should only ask questions that is they asked their mum she would have to tell them the truth.

- □ *Stop fishing for compliments.* Don't ask questions such as 'We have this idea for an app, do you like it?'. Try as much as possible to not talk about your idea and instead focus on the customer and their needs. For example, you can say: 'Talk me through the last time you had this problem'.

- □ *Stop pitching.* Don't describe the features of your intended solution. We are not making a sales call. We are trying to learn about our customers. So instead you can ask customer to describe the biggest challenges they face when they are trying to get their jobs done.

- □ *Don't ask questions about the future.* Questions such as 'Would you buy this product?' or 'How much would you pay for this?' are not very useful. People can't predict what they will do in the future. If you really want to know whether people will buy something, it is better to run experiments that measure actual behaviour (see Chapter 4).

---

[21] Fitzpatrick, R. (2014). *The mom test: how to talk to customers and learn if your business is a good idea when everyone is lying to you.* Amazon CreateSpace.

3.9 Empathy map

Now that you have been out of the building observing and talking to customer, it is time to create a picture of what we have learned. Furthermore, there could be members of our team who were not out of the building with us, who also need to develop an understanding of what we learned. A good tool for developing this picture is the empathy map. It allows teams to map what they saw customer say, do, feel and think.[22]

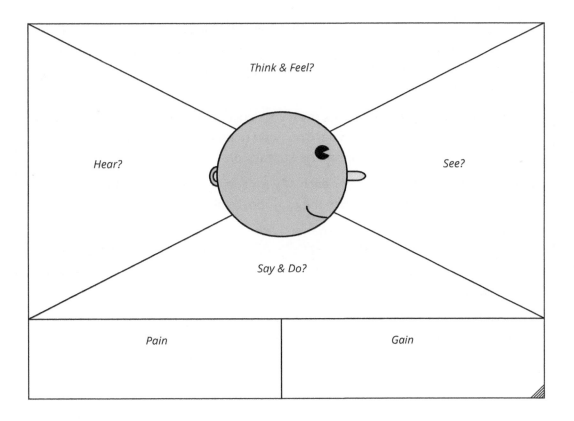

[22] Gray, D., Brown, S. and Macanufo, J. (2010). *Gamestorming: A playbook for innovators, rulebreakers, and changemakers.* Sebastapol: O'Reilly Media.

How to do it

Before you start the session, make sure that all the key members of your team are present. Developing a point of view about our customers is something that is for every member of the team not just the product developers. So, make sure a cross-functional group of people is present (e.g. sales, finance, marketing, technology and customer support). Also make sure you have the necessary tools for the session; a large empathy map, sticky notes and sharpies.

1. The first step is to unpack all the research work that you did when you were out of building. Capture your main observations and key themes. Share with the teams some stories from your experiences with customers. Ask them what stands out or surprises them. Have them capture the key observations and themes from your story.

2. After this, identify key segments from your customers. Which individuals share pains, gains and jobs to be done? You need to identify these segments because you will create an empathy for each one.

3. Now complete an empathy for each of your customer segments. Write one insight per sticky note. What do they say, do, think and feel?

4. When you are done mapping, analyse each map. What stands out to you? Are they any contradictions? Do people say one thing and do another? Why might that be? Are there any other insights from your research you can bring into the conversation?

5. As you do the analysis, also look for similarities between customer segments. This may help with identifying needs that are shared by a broader group of customers.

Q Tips

When working on your empathy maps, here are some useful tips you can follow.

- Use sticky notes, do not write directly on the map.

- Create different maps for different customer segments. However, if you use one map, then use different-coloured sticky notes for each customer segment.

- Focus on what you learned about the customers and their needs. This is not the time to start brainstorming solutions.

- At the end of the session, make sure there is a shared understanding of the empathy maps throughout your team.

3.10 Update the circle

At this point, we need to revisit the customer profile on our value proposition canvas. Remember that all the work we have been doing so far exploring customer needs was triggered by the risky assumptions we identified when we reviewed our value proposition canvas. So, the question we need answer is whether all the lessons we are learning support our assumptions and hypotheses.

- *Customer jobs, pains and gains:* As team, we need to review each sticky note on our value proposition canvas. For each sticky note, we need to discuss whether what we learned from customers supports that assumption. Were we correct in making that assumption about customer jobs, pains or gains? Do we need to revise the assumption? Or should we remove it from the canvas altogether?

☐ *New learnings:* We will also have learned new things about our customers that were not part of our original canvas. These new learnings will need to be added to the canvas as well. Remember to put only one job, pain or gain per sticky note.

☐ *Make decisions:* The more we get out of the building, the more our value proposition canvas will change. At some point, we will have to make decisions. Do we know enough about customers to start thinking about solutions? Have we found meaningful customer jobs to be done? How strong are the customer pains/gains? How are they currently solving their problems or meeting their needs? Do we think customer may pay for a solution to help with their jobs to be done?

The answers to these questions will inform the decision to stop, pivot or persevere. But once we have a sense that there is a real need that we can serve, it is time to revisit the value map on our value proposition canvas.

3.11 Update the square

As our customer profile has been evolving based on customer learnings, we now need to revisit our value map. It is highly like that we will have to update our value map based on learnings. Remember that our goal is to get to problem-solution fit. This means that we need the features of our product or service to match our customer jobs to be done, pain and gains.

☐ *Product and services, pains relievers and gain creators.* Given what we now know about customer jobs, pains and gains, do we still have the right product or service in mind? Do we have the right feature set to relieve the pains or create the gains that customers want?

☐ *Learn then confirm*: Innovators have a tendency to take customers at their word. This is partly because of the confirmation bias we spoke about earlier. It is important to recognise that, even if our revised value map is based on customer learnings, our ideas about solutions are still untested. So, an important principle for all innovators to follow is to *learn then confirm*. In other words, talk to customers then get them to do something.

- *Identify assumptions:* So, after we finish revising our value map, we must resist the urge to make the leap to building a product. Instead, we have to review the map for risky assumptions. Are we confident that customers would want the solution in the way we are imagining it? Will they be willing to sign up for updates, pre-order, engage in a free-trial or make a purchase?

- *Early solution testing:* Believe it or not, these questions can be answered without building a product. Furthermore, the questions can be answered, not with interviews, but by running experiments that get customers to do something. We will now turn to a few examples of experiments we can run. Such early solution testing, ensures that we enter the Validate stage of the Lean PLC with clearer ideas about the solution we should be building.

3.12 The landing page

The landing page is a single web page that is used to test our value proposition. You can also use flyers and posters to run similar tests. The goal of the test is to check whether our value proposition resonates with customers. Our landing page simply says what the product does and then asks customers to perform some sort of action that registers their interest.

How to do it

Below are some elements of good landing pages.

- *Headline*: At the top of the page make sure you have a clear headline of what your product is. It is good at this point to have a name for your product. Although, you can actually use the landing page to test out different product names too! A nice image on the landing page also helps, but this is not absolutely necessary.

Figure 3.5

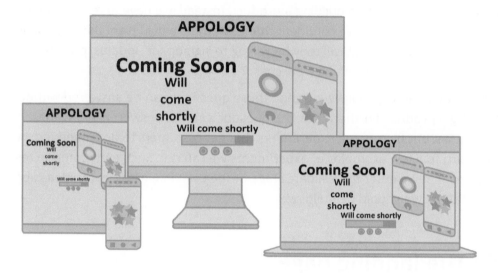

- *Value proposition*: Underneath the headline, you can put a clear value proposition. This is not a description of the features of your product but a statement or promise of the value to be delivered to customers. For example, Lyft once had a landing page with a value proposition statement that simply read, 'Make up to $35/hr driving your own car'.

- *Call to action*: This is the best way to gauge customer interest and intent. After seeing the headline and value proposition, you have to get them to do something. A call to action can be a button that customers can click to sign-up for updates, sign-up for a free trial, pre-order or buy the product. The conversion rates on your call to action is the real data of interest from a landing page. This shows us whether we are thinking of the right solution.

- *A/B testing*: A single landing page can be useful for testing your value proposition, but this process is more powerful when you test different versions of your landing page and compare the results. For example, you can randomly allocate people to see different version of the page with different value propositions or product names and see which one resonates the most.

3.13 Comprehension test

The results obtained from a landing page or a test advertising campaign are only meaningful to the extent that people understand the headline or the value proposition. It is possible for an experiment to fail simply because people did not understand the value that was offer. As such, before setting up your landing page or running a test adds, it makes sense to run a comprehension test.

How to do it

This test takes about 15 minutes to complete. All you need to do is to find a sample of early adopters. The test must be run one person at a time:

☐ Create the landing page or simply write value proposition statement clearly on a piece of paper.

☐ Show the landing page or value proposition statement for a few seconds. Provide enough time for them to read it (e.g. 5–10 seconds).

☐ Remove the landing page or value proposition statement and ask the person to tell you what the offering is in their own words.

☐ Capture their responses and compare to your own landing page or value proposition statement.

If what people say is similar to your value proposition, then you have a clear offering. However, if people are saying things that are different from your value proposition then you might have to revise your statement to make it clearer.

Q Tip

Remember that if you revise your value proposition statement, you have to retest the new statement with customers.

3.14 The explainer video

Another way to ensure that people understand the value proposition is to create an explainer video. This is especially useful when you are thinking of creating a complex product or a product that serves complex needs. A popular example of an explainer video is the one Dropbox created before they had finished building their product. This video illustrated how the product 'worked' and resulted in over 50,000 signups for a product that was yet to be completed!

How to do it

With explainer videos it's all about the script. This is something that has to be well designed. A good script for an explainer video has the following elements.

☐ Start with the problem or customer need to be solved.

☐ Then present current solutions and why they are not providing value.

Figure 3.6

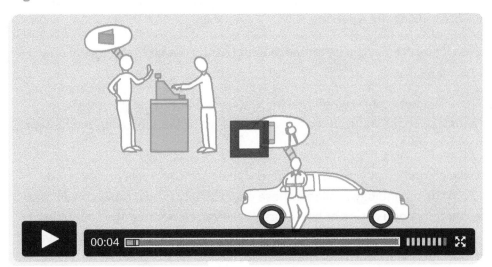

☐ Now introduce your solution and describe how it provides value.

☐ Provide some social proof that your solution works.

☐ End the video with a call to action.

> ## 🔍 Tips
>
> ☐ Given people's attention spans, short videos are better than long ones. You also have to get people's attention early, so make the first 20 seconds count.
>
> ☐ But remember that short is not the same as fast. So, don't jam your scripts with loads of information and have the narrator talk fast just so that you keep the video short. Instead, curate your script to ensure that only the essential information and details are provided.
>
> ☐ Even if you are thinking of a complex solution, try and keep it simple. Avoid jargon and use the everyday terms that customers use to describe their problems and needs. This means we have to pay attention to the audience we are making the video for.
>
> ☐ Even if you are using a script make sure that the video feels natural and unscripted.
>
> ☐ There is a temptation when making videos to want to highlight the cool features of our product. Please avoid this and instead focus on the benefits to the customer.
>
> ☐ Make a good quality video with good visuals, picture quality and sound. Given the technology available these days, quality is now table stakes. Use animation, good actors or narrators. However, remember that the ultimate measure of quality is how easy it is for customers to understand your value proposition after watching the video.
>
> ☐ Explainer videos are usually part of a landing page. As with all landing pages, always include a call to action.

3.15 Paper prototypes and storyboards

If making a video is too costly an option for you, there is another way to test your value proposition visually. With paper prototypes, you can talk people through the potential experience of using your product. You can then invite them to give you feedback about your idea. You can even allow customers to scribble on the paper prototype. They can then show you the places they may look for a feature and click to find value. This feedback is powerful as an early indicator that you are headed in the right direction.

Storyboards are also a great visual way to test your value proposition. You can sketch out a narrative storyboard or a short cartoon that tells the story of how your customers can use your product to solve their problems. Such a visual presentation can help in two ways: it can help you check whether your understanding of the customer context and needs is correct; it can also help you test whether your proposed solution resonates with customers.

As shown in the image, a storyboard needs to have only a few panels that show:

- the problem you are trying to solve for customers
- the context in which customers will encounter and use your solution

Figure 3.7

1

User arrives at station, sees kiosk and long queues for the ticket office

2

User goes through monitor

3

On payment screen, he chooses mobile phone

4

He scans the QR code

5

Follow instructions to pay and have tickets delivered

6

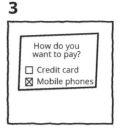

Gets ticket delivered to his phone. Speedy!

□ how the solution will help them solve their problem or get their jobs done

□ the key benefits they will get from using your solution.

Q Tips

□ Use your prototype or storyboard is created to test specific assumptions.

□ Use these tools test whether the customer need is serious enough and also whether the solution you are thinking of is the right one.

□ Keep the visuals simple and easy to understand.

□ Do not pitch or try to sell your idea. Instead allow customer to review and critique the paper prototype or storyboard.

□ Allow customer to lead the conversation. Try not to guide them too much.

□ Make copies of the storyboard and allow customers to write their feedback, preferred solutions and workflows on the map.

□ Use the lessons learned to iterate on your solution ideas.

3.16 Customer journey mapping

A variation of the storyboard is the customer journey map. This map is a graphical representation of the customer's experience via key touch points with your organisation and the solution. This process is often used for products that are already in the market. However, it can also be used for new ideas as a way to test our thinking about how our solution is going work within a customer journey.

How to do it

- [ ] Clarify the customer segment for whom you are mapping the journey.

- [ ] Also, be clear about which journey to are mapping (e.g. awareness, purchase and first experience).

- [ ] Walk a mile in your customer's shoes. Imagine all the touchpoints that they will have with your organisation and the solution.

- [ ] Capture the activities and feelings that should be part of each key step along the journey.

- [ ] Make clear the multiple ways that customers can reach the same touchpoint (e.g. customer can get to the purchase point via sales calls, website or visiting a physical store).

- [ ] Differentiate touch points that are visible to the customer, versus those that happen backstage within your company.

- [ ] Make your journey easy to understand. You can use A3 or A4 size paper in landscape. Map the steps of the journey horizontally along the top. Along the vertical axis, you can map the customer activities, thoughts and feelings that are part of each step.

- [ ] Test the journey map with customers. Ask them to identify additional steps and activities. You can also ask them to rate or rank the steps that are most important to them (e.g. on a scale of 1–10).

Q Tip

Use customer feedback to identify any gaps in understanding their journey that you may have.

Figure 3.8

3.17 Picnic in the graveyard

When to use it

This method is called picnic in the graveyard, because it focuses on examining products that have failed and 'died'. The goal of all the methods and tools we have described so far is to help us get an idea of the product we should be building. What features should it have? How can we deliver value? What is the minimum set of features that can be in our first version of the product? However, even if our product idea is highly creative and original, there is nothing new under sun. It is likely that someone else has had the idea before. The picnic in the graveyard[23] method is based on finding products similar to our idea that been launched and failed in the market.[24]

Figure 3.9

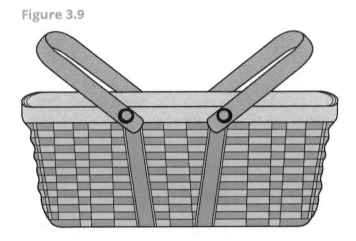

[23] Murphy, S.K. (2012). Pretotyping – Techniques for building the right product. Available at http://www.skmurphy. com/blog/2012/03/06/pretotyping-techniques-for-building-the-right-product/

[24] Kromer, T. (2015). Generative research: Picnic in the graveyard. Available at https://grasshopperherder.com/ generative-research-picnic-graveyard/

How to do it

☐ *Explore* – Take time to look for products that have been launched and failed. Desk research is good for this, but you can also speak with colleagues and customers. In fact, you can explore some of the solutions that customers are currently using that are not delivering value.

☐ *Research* – Talk to customers to learn about what they disliked about the failed product. You can also reach out to the companies or founders that launched the failed product and learn from them. Some founders may not be willing to talk, but you will find that most founders are willing to help.

☐ *Analyse* – Review your findings for hints on how you could make the products better. Many products that fail are usually close to the right solution but are missing one or two features.

🔍 Tip

If you are going to be introducing a product that is similar to those that have gone before, then you need to be sure about your key differentiator.

3.18 The market opportunity

As the picture of your target customer and their needs becomes clearer, it is time face up to a very important question: does your target customer represent a market that is large enough to build a sustainable business? This is an important question because finding a group of customers with a need is not the same as finding a profitable market. As such, you can use the desk research techniques we described earlier to perform some market analysis.

How to do it

1. An important first step in your market analysis is to consider the total size of your market. When doing this analysis, you must consider both volume and value. Volume speaks to the potential number of customers, while value speaks the total amount of dollars available in the market. Ideally, you want a market with high volume and high value. However, it is likely that you might have to make trade-offs.

2. Next, you need to be more specific about the type of customer within that broader market that you are going to target with your product. We often encounter innovators who think that they their product is for everyone. They commonly make broad statements such as 'My product is for everyone who needs X'. No product is for *everyone,* even if they share the same needs! As such, we have to use our learnings to select a more specific target market.

3. Another mistake innovators make is to pick a random percentage and use that as their target. For example, people will say things like, 'There are over five million dog owners in England, and if we get just 10 per cent of that market then we will be rich!' Such estimates do not provide a proper basis for decision making. Instead, we need to define a specific target market and then estimate what size of the total market those people represent.

☐ One way to make these distinctions is to calculate TAM, SAM and SOM. The total addressable market (TAM) is the total market demand for a product. The serviceable addressable market (SAM) is the part of the TAM that is served by your product or service. Finally, the serviceable obtainable market (SOM) is the part of the SAM that you can realistically reach.

☐ In addition to the potential size of the market, you also have to consider competitors. Are there any strong competitors in the market? What is their market share? Is this a *red ocean* where you have to succeed by taking customers away from the competitors? If yes, what are the switching costs for your customers? Or is there a market segment that competitors have ignored that will be served by your product? After analysing competitors, you may need to revise your market size numbers, especially for SAM and SOM.

☐ Finally, you need to analyse whether there are any barriers or regulations that make it hard to enter your target market. For example, you may need licences and permissions from other companies to use certain technologies. There could also be barriers that restrict access to distribution channels (e.g. exclusive contracts). Barriers often raise the costs of entering a market, which must then be offset against the value of that market.

After completing your market analysis, you should then return to question; does your target customer represent a market that is large enough to build a sustainable business? If the answer is yes, then you are good to go. If the answer is no, you can make the decision to pivot to another target customers or decide to stop the project entirely.

🔍 Tip

When performing market analysis, please be honest with yourself. Non-existent markets won't just pop up after you have created the product. Remember that people will not pay to solve a problem they don't have!

3.19 On surveys and focus groups

So now it's time to address the elephant in the room. You may have noticed that our review of methods left out surveys and focus groups. Although these two methods are quite popular among market researchers, there are not top of the list for us in the Lean PLC. Our concern is with getting a deep understanding of customers and their needs. We feel that these two methods are not best suited for doing that.

Surveys do have their place in the research methods toolbox, but only if they are used with other methods. The reason we don't encourage innovators to use online or mail survey is because violates the 'actual place actual thing' rule (i.e. Genchi Genbutsu). When trying to learn about customer there is nothing better than going to see their lives for yourself. Surveys may be useful when you have a product out in the market. But even then, these market data must be supported by going to the market and speaking to real people.

Focus groups are also a popular market research tool. However, this method of engaging with customers increases the social desirability challenges we highlighted earlier. People's behaviour is strongly impacted by the presence of others. It's always important to remember that we are not just interested in hearing customers speak. We are interested in what they *really* think and feel. So, it is important to put customers in situations where they will be comfortable sharing their inner thoughts with us. Focus groups with strangers are not really the place for that.

Q Top tip

We are not saying that people should never use surveys or focus groups. However, these methods should be used with caution and they should be used as part of a larger toolbox of research methods (i.e. learn then confirm).

3.20 Making decisions

We have reached the appropriate point in our journey to remind ourselves why we are doing all this work in the first place. During Explore, the methods and tools we have described are used to test some key assumptions: is there a real customer need, how strong is that need, what are the customer jobs to be done and what might be good solutions to meet those customer needs? At the end of every experiment, we have to review our learnings and make decisions.

> *Remember that if we are making something that nobody wants, then it doesn't matter if we are on time and on budget.* – Eric Ries

Our main goal as we move through the Lean PLC is to make products people want, and to figure out a profitable business model with which to deliver that value. While we do our work, we will run down rabbit holes and dead ends. As we pivot and iterate our way through Explore, we will eventually come to the end of our runway. Then it's time to face reality:

> *Have we found a real customer need and potential solutions for that need?*

If the answer to this question is 'YES', then we are ready to start building a solution. However, if the answer is 'NO', then we have to make a decision. Do we pivot or do we stop? This decision is a judgement call based on learnings, context and available resources. It is possible that while you have been out of the building, you have picked up a strong signal of a potentially strong need or new customer segment to pursue. In this case, if you have the funding and resources you can remain in Explore and make a pivot.

On the other hand, it is sometimes a good idea to stop. This is particularly important, if there is no real customer need. Rather than waste time and money insisting on pursuing a dead end, it is better to save our resources and deploy them in new ideas. We may not throw away our idea. We may just park it for a while. It is possible that overtime, new insights will emerge within our company that can revive our idea in some form. However, for the time being we have to STOP.

Figure 3.10

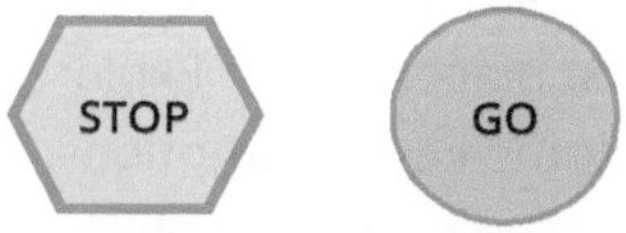

STOP   GO

3.21 The business model

If we have found customers with a real need and we have an idea of the solution we might want to build for them, then it is time for us to start thinking about the business model. Beyond ideation about products and services, business model design allows us to take a more holistic approach to innovation. We can see how our ideas about products and services work in concert with the key partners we might need to deliver value, the channels or customer relationships we might need, potential cost structures and revenue models.

One of our favourite tools for business model design is the business model canvas. Developed by Alexander Osterwalder and colleagues,[25] it has nine key elements that make up typical business models. Each of these elements is described below.

Figure 3.11

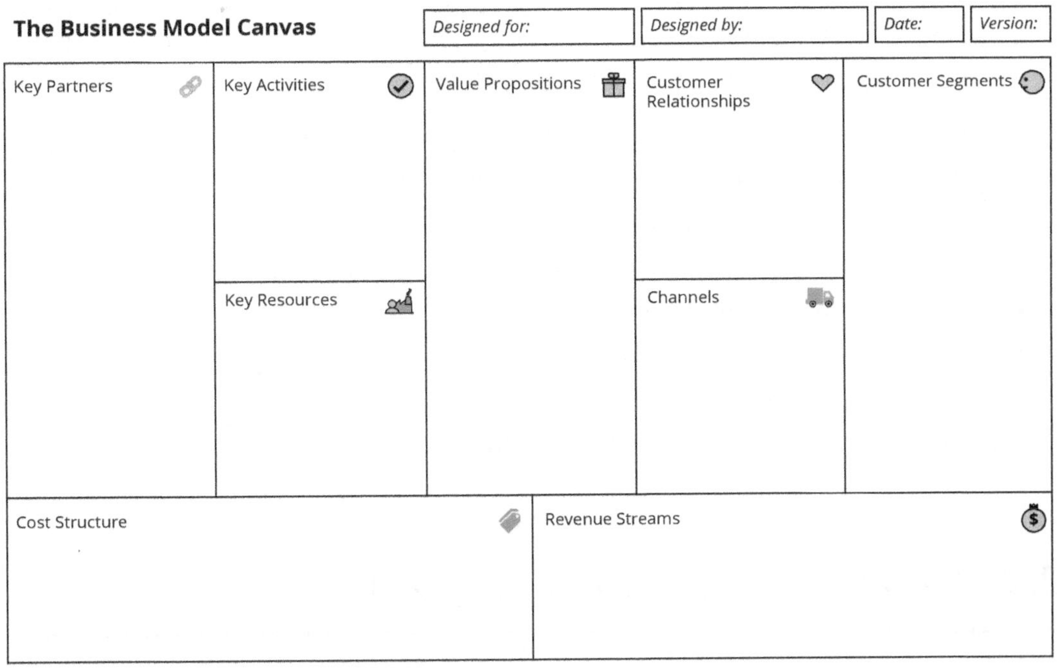

The Business Model Canvas

| Designed for: | Designed by: | Date: | Version: |

| Key Partners | Key Activities | Value Propositions | Customer Relationships | Customer Segments |

Key Resources — Channels

Cost Structure — Revenue Streams

25 Osterwalder, A. and Pigneur, Y. (2010). *Business model generation: a handbook for visionaries, game changers, and challengers.* New York: John Wiley & Sons.

- *Key partners*: It is sometimes important for companies to form partnerships with other organisations. This can help your company reduce risks and acquire capabilities it does not currently have.

- *Key activities*: In order to make the business model work, there are certain activities your team will have to do well. These can include product, problem solving, marketing or developing a network/platform.

- *Key resources*: This refers to the physical, human, intellectual and financial resources you will need to create value and make your business model work.

- *Value propositions*: This describes the value that you provide your chosen customer segments. This is a summary of the detailed information you have already captured in the Value Proposition Canvas.

- *Customer relationships*: This is about the type of relationship you are going to have with you customer segment. There are various types of relationships including dedicated personal assistance, self-service and automated service.

- *Channels*: This is how you are going to reach your customers. How will they learn about the value you offer? How will you communicate with them? And how will they get the product or service delivered to them?

- *Customer segments*: These are the various groups that you have identified during Explore that will be served by your product or service.

- *Cost structure*: This describes all the fixed and variable costs you will need to create and deliver value. Identifying risks around costs is an important part of deciding whether you have a profitable business model.

- *Revenue streams*: This describes how your organisation is going to capture value back from the customers. There are various revenue models you can have including asset sale, usage fees, subscription fees, renting and licensing.

3.22 Business model prototyping

The business model canvas can be used as a prototyping tool. This business model design method allows us to think through various types of models for our product or service. As we start to settle on the right solution for our customers, we need to imagine several business models we can use to create and deliver value. Prototyping prevents us from getting trapped in the local maxima of using the first business model we can think of. Below is a facilitator's guide on how to run a business model prototyping session.

How to do it

1. Gather a small cross-functional team of 4–6 people in a room. Make sure you have an A0 size business model canvas up on the wall, loads of square sticky notes and sharpie pens.

2. Before starting the session, you can give the team the following business model design rules:

   (a) Do not write on the canvas.
   (b) Write on the sticky notes and place them on the canvas.
   (c) Put only one idea per sticky note.
   (d) There is no leader of the canvas, every team member should be allowed to write and contribute.
   (e) Focus on quantity over quality. We will review the work later.

3. Before starting the actual design session. Do a warm up exercise. For example, ask the team to build a business model for a cow or packet of gum.

4. When you are ready, have the team map their first business model. They can complete each section of the canvas using the sticky note. Encourage them to think through all aspects of their business model including key partners, key activities and key resources.

5. When this first version is complete, the team can take a picture of this model using their smartphones and store it in the 'fridge' for future reference.

6. Now that we have our first business model, we can start to do our prototyping. In order for this to work, we ask the team to imagine a series of 'what-if'. scenarios. Using their current business model as a starting point, you can ask the team to imagine how they would change or adjust the model in responses to a few scenarios. Below are a few examples as a guide, but you can come up with your own.

(a) Imagine you couldn't use the Internet or a mobile application to sell or market your product. How would this change your business model?

(b) Whatever your current price levels, multiply them by ten. How would you improve your value proposition to justify the cost?

(c) Give your core value proposition away for free. How are you going to make money using this business model?

(d) How does your business model change if your costs go up by 60 per cent?

(e) How does your business model change if a competitor comes in and takes over 50 per cent of the market share?

7. Have your team map 3–5 business models based on the above 'what-if' scenarios. Timebox the activity by giving the team 15–20 mins for each mapping iteration. For each business model they map make sure they take a picture of it when they have done.

8. At the end of this divergent thinking, it is time for the team to converge on one model. Pull out the business model that the team began with. Given everything they have designed so far, ask them how they might want to revise their initial business model. Allow the 30–60 mins for them to design a final business model, which will form the basis of our work going forward.

3.23 Identify assumptions – make plans to test

At this point we have achieved three key milestones. We have discovered a real customer need that we can help solve, we have identified a value proposition and potential solution that resonates with customers and we have designed our 'Plan A' business model. Indeed, the lessons we have learned so far will have informed our business model design. However, all this does not mean that we are ready to build and launch our product. There are still many assumptions about the solution and business model that need to be tested.

Our first business model is referred to as our Plan A for reason. We will have to iterate our way to Plan B, C, D. . . Z using lean innovation methods. So, as we did with our value proposition canvas, we need to identify our risky assumptions (see Figure 2.2).

☐ As such, the team must step back after mapping their business model and review each sticky note using the same key question: Do we have evidence for this or is it an assumption? When in doubt, it is an assumption and we must put a large red X on all sticky notes with untested assumptions.

☐ After this, we will then rate each untested assumption in terms of how critical it is to the success of the business model (1 = *Not critical* and 10 = *Critical*). Remember that an assumption is critical if you have to stop the project if it turns out you are wrong.

☐ After identifying our riskiest assumptions, we can then make plans to test to them. The next stage of the Lean PLC is the Validate stage. During this stage you will test your assumption about the solution as you are creating it. You will also test your assumptions about the business model. Your team can use the experiment canvas to start coming up with different ideas for the experiments you might run to test your ideas (see Figure 3.1). This is the critical final step as you prepare for the Validate stage.

3.24 Preparing for Validate

If you have done the activities we described above, your team is ready to ask for a larger investment to move to the Validate stage. Indeed, the product council is happy to invest a larger amount of money in your project because you have validated real customer needs and a resonant value proposition. However, Lean PLC rules still apply; do the right things at the right time. At this stage, teams should not be asking for massive investments to launch their product at scale. Instead, they should be asking for the minimal investment they need to test their assumptions around the solution and business model. We may include:

- brief details about the solution we plan to work on and how the solution supports the strategic goals of our company

- an update of the work we have done so far and how the evidence supports our proposed solution

- an update of our research on the market opportunity and whether it represents a real market

- our assumptions about the right product or service to create for customers

- our assumptions about the right business model to create and deliver value

- the assumptions we plan to test during Validate, and brief details of the tests we plan to run

- the resources and funding we think we will need to complete the iterative creative of the solution and validate our business model.

We are still not looking for a 40-page business case with five-year projections in it. The product council or investment board is looking for evidence of a real customer need that faces a large enough market and clear next steps that take us closer to a successful launch.

SUBMISSION TEMPLATE
Explore to Validate SUBMISSION

Idea ownership

Investment board	
Business sponsor	
Product owner	

Idea overview

Product name	
Idea description	
Strategic fit	

Target customers and jobs to be done

Describe your potential customer segments and their needs or jobs to be done.

Customer segments	Jobs to be done

Explore lessons learned

Provide an overview of the assumptions you tested in Explore and the lessons learned. Add more customer assumptions as necessary.

We believe that:	
To verify that we:	
And we measure:	
We learned that:[25a]	

[25a] Osterwalder, A., Pigneur, Y., Bernarda, G. and Smith, A. (2015). *Value proposition design*. New York: John Wiley.

Unexpected learnings

Provide a summary of the unexpected learnings you got during Explore.

Market opportunity

Please update your learnings about the market opportunity below:

Total addressable market (TAM) in dollars ($)	Serviceable addressable market (SAM) as % of TAM	Serviceable obtainable market (SOM) as % of SAM

Market description

How much did you spend in explore

Provide a breakdown of your spend during Explore.

What are your plans for validate?

Please provide a summary of the activity you plan to do next.

Assumptions to test during validate

Provide an overview of the assumptions about the business model you plan to test during Validate, how you will test them and your success criteria.

We believe that:	
To verify that we:	
And we will measure:	
We will know we are right if:	

Market opportunity

Beyond the research that you have done so far, please provide an overview of any further research on the market opportunity that you plan to do in the next phase.

Resources and funding requested

To complete the Validate stage, we are asking for (e.g. **dollars, time, people**):

* Please note that you also have the option to remain in Explore, go back to Idea or stop the project entirely. If either of these options are what you are choosing, then you need to adapt this template so that you can update the product council on work done so far, key lessons learned, what you plan to do next and the resources you need

An interview with:
The Innovaid team at Rabobank

Rabobank is a Dutch multinational bank headquartered in Utrecht, Netherlands. It has over 40 000 employees and is among one of the largest financial institutions in the world. The bank is a global leader in food and agriculture financing. This interview case study was conducted with Rabobank's Innovaid team. This team is led by Siddi Wouters, Global Head of Innovation at Rabobank. Innovaid is responsible for the education and acceleration of innovation within Rabobank.

In your experience, what are some of the top challenges large companies face when it comes to testing new ideas for products or services?
There are a couple of challenges large organisations face when it comes to testing new products/services. Most organisations follow the horizon classification of McKinsey: Horizon 1 and 2 innovations are implemented by the Execution part of the company and Horizon 3 initiatives are created by a separate department, accelerator or incubator. The biggest challenge with this approach is that the learning curve, mindset and skillsets gained grow faster in this separate unit than within the execution (main) part of the organisation. At Rabobank we help bridge that gap by facilitating knowledge and coaching from our innovation unit (the separate department) to include the entire organisation no matter what type of innovation initiative you want to investigate. Getting the business lines to learn as fast as the innovation teams do, is one of our goals. We are still trying to balance the often reactive and time-consuming demand from the organisation and our own mission to create new, Horizon 3 initiatives.

Another challenge within large organisations is the time provided to employees to 'work on innovation'. Often, employees do not get dedicated time to work on innovation, as there is no time, mandate or priority to focus on innovation. Which leaves employees in many situations demotivated to really get the idea going. At Rabobank we enable dedicated time to work on innovation through campaigns and innovation programs to have a continuous impact on employees, ideas and the business as a whole.

Finally, the entrepreneurial mindset, to find ways to run experiments, is limited or less appreciated within large organisations as it is not a mandatory integrated part of the day-to-day activities and can be viewed as risky as it can affect business. Innovaid aims to change this.

As the Innovaid Team, what best practices do you recommend companies like Rabobank put in place to help their teams generate great ideas?

The Innovaid team is responsible for empowering Rabobank to innovate more, faster, fact-based and customer oriented. We do this by training, coaching and facilitating the time, room and space to innovate. Our current focus and starting point is on idea generation, for which we created our own Ideation game and workshops. We train our enthusiastic employees how to facilitate their own workshops and campaign globally to create traction. This way we spread innovation through the bank exponentially.

Without giving away too much detail, can you share a story of a team that you have worked with in Rabobank that did great work of testing their idea? What did they do well?

We can name many teams that tested their ideas successfully. However, teams that potentially did even a better job, are those that killed their initiative based on customer feedback. These 'successful failures' learned what won't work and acted upon that instead of continuing based on their gut feelings. The teams we work with have one objective: understand your customer. We guide teams through several phases: Ideation, Problem-, Solution- and Market fit. All these phases have different aspects and deliverables, but are all focused on one thing: solving a problem for the customer! When the results or assumptions don't match we take a step back to align or even stop with the initiative. When they do align, we coach them through the next phase.

Finally, what do you think companies should do once they have found an idea that has potential traction?

If homegrown initiatives show positive signs of traction, companies should unleash the teams. Depending on what the team needs to grow, from a staffing point of view, as well as from a marketing point of view, the company should reward the team.

What companies should avoid is funding the team for multiple years in a row, but instead offer short funding cycles that are effective and result orientated: are we solving a customer problem?

4

Validate

4.1 Welcome to Validate

Welcome to Validate! In Explore you were successful in validating customer needs and the value proposition that resonates with them. It is during Validate that you will build your product or service and find the right business model. You will now need a larger team than in Explore, that is capable of iteratively building your product and testing the business model (e.g. a product manager, engineers, coders, designers, a UX expert, a marketing expert with some support from sales). The Validate stage can be split into four key phases.

1. *Validate solution*: This phase focuses on creating the right solution for our customers. We will start with a minimum version of our product. We will then use customer feedback to iteratively make the product better.

2. *Validate business model*: As we build our solution, we will also use the minimum viable product to test other aspects of our business model. We will examine question such as: What is the right price for our product? What is the right channel to deliver value? Which key partners do we need to work with?

3. *Identify growth engines*: While we are validating the business model, we also have to reconfirm the market opportunity and competitive landscape. We also need to start to thinking about the right engine of growth for our product. How do we acquire more customers to grow revenue and profits?

4. *Prepare for growth*: At the end of Validate, you should have the right solution and a validated business model. In Grow, you will need to fully launch your product, and take it to scale. So, at the end of Validate you need to state clear growth hypotheses and define the resources you will need to launch your product and take it to scale.

4.2 The minimum – the viable

The key activity during Validate is building our solution and using that process to test our assumptions about the business model. Even if we have a clear idea of what we might want to build, the key discipline here is to avoid building the entire solution in our silo without getting customer feedback. Instead, we must begin with a minimum viable version of our intended product and then iterate our way towards the full product using customer feedback.

The idea of a minimum viable solution often raises questions among innovators: What is minimum? What is viable? We have worked with teams that argue about the level of fidelity the first version of their product should have; and which features to include or exclude. One way to resolve this debate is to remember the basic philosophy of lean thinking. All the tools and methods that are described in this book and others are not used for their own sake. Instead, they have to be viewed as instruments for testing our business model assumptions.

Remember that the final exercise we did during Explore was to map our business model and identify our key assumptions. It is these assumptions that we must use to decide what is minimum and viable. Our decisions depend on what we are trying to learn and the assumptions we plan to test. In that regard, *minimum* refers to the smallest thing we can do tomorrow to start turning our assumptions into knowledge. *Viability* is then judged by the ability of our chosen method to provide us with the answers we need.

At the beginning of Validate, we are trying to test our solution hypotheses. This raises technical questions about whether we will be able to create the solution we are thinking about (i.e. can it be done?). It also raises questions about what form the solution should take and whether that will meet customer needs (i.e. how should it be done?). Later on, we will use the solution we are creating to test other aspects of our business model (e.g. pricing). Our first choice concerns which set of assumptions we want to start testing. This choice is what we will then use to inform what goes into the first minimum viable version of our solution.

Stated in another way, the minimum viable product is simply an instrument for running experiments. Our job as innovators is to make sure we include in our solution the features that are necessary for testing our assumptions and no more. This discipline to avoid doing more than necessary is hard for innovators because we always want to put our best step forward and show customers beautiful solutions. But if we change our definition of 'best foot forward' into the thing that will most help us to learn, this might make it easier for us. It might be comforting to remember that later on during the Grow stage, our 'best foot forward' will be the thing that will most help us to scale.

4.3 Measuring the right things

In addition to designing the right experiments, we must also make sure we are measuring the right things. As innovators, we are often keen to find out how much people like our ideas. 'They love it!', we often hear proud innovators say. But finding a few people that like our product is not the same thing as finding a market. In fact, finding a million people that like our product is not the same thing as finding a profitable business model.

There is a difference between vanity metrics and actionable metrics. *Vanity metrics* make us feel good about our product (e.g. number of likes on our Facebook page or number of views on our website). However, such metrics do not help us make decisions about our business model. In contrast, *Actionable metrics* allow us to make decisions about our business model and what to do next. This begins with tracking business model relevant metrics. What will it look like if our product succeeds? How will our customers be behaving? What is a good retention rate? How will we make money? How will we know we have reached the right price point?

Actionable metrics speak to the reality of what might make our business model fail. They are based on our key assumptions. Actionable metrics are also comparative in nature; meaning that we can take a baseline measure and then track changes over time. This is why ratios such as percentages make great actionable metrics because they allow us to compare data across different cohorts and time periods. In contrast, cumulative metrics, such as total number of customers, are always going up into the right. Such metrics are not useful for comparing performance day to day, week to week or month to month.

We will illustrate key metrics as we go through this chapter. But for now, we will turn to the methods and techniques we can use to test our assumptions during Validate.

4.4 Prototyping

Prototyping is a good way to begin testing our assumptions about a solution. Prototypes can be used to test technical risks about our ability to create the solution.

They can also be used to test whether the solution we are thinking of creating is the right one for customers. The choice of prototype depends on the solution we are thinking of creating and assumptions we are trying to test. We can create digital prototypes and physical prototypes.

> *It's fine for a prototype to have issues and be ugly. The goal of a prototype is to evaluate an idea. Once the idea is validated, you will iterate to improve the prototype and make it an app that can be shipped. – Olivier Collet[27]*

How to do it

☐ *Digital prototypes*: For teams creating software products, there is often a temptation to just start coding. However, before we start building our product we can use prototypes to learn which features resonate with customers. Digital prototypes can range from low fidelity wireframes to high fidelity software mock-ups. You can tailor the fidelity of your prototype to the target audience. We recommend beginning with a low fidelity prototype, presenting that to customers for feedback, and then using that feedback to increase the fidelity of our solution. Digital prototypes are particularly powerful for testing the user flow in our software.

☐ *Physical prototypes*: Even when making physical products, prototypes are a good place to start. In particular prototyping can help a company figure out whether they have the capability to create a product. It allows us to learn early whether they are any challenges in our manufacturing process that we need to deal with before we scale. Physical prototypes also allow us to test usability and customer preferences. With the advent of 3D printing, the process of creating physical prototypes need not be expensive.

---

[27] Grothaus, M. (2015). *How to go from idea to prototype in one day.* Fast Company. Available at https://www.fastcompany.com/3045934/how-to-go-from-idea-to-prototype-in-one-day

> ## Q Tips
>
> ☐ Always test your prototypes with real customers from your target segment. Do not test on colleagues from work or members of your team!
>
> ☐ Remember to prototype only what you need to test your assumptions.
>
> ☐ Don't strive for perfection. Be comfortable sharing rough version of the product if this helps you learn what to do next.
>
> ☐ When making digital prototypes, avoid using 'lorem ipsum' text in your mock-ups. It is better to mock up some 'real' text to present to customers.

4.5 Concierge minimum viable product (or Concierge MVP)

Even when the intention is to build a digital product that delivers value in an automated fashion, it is possible to begin the process of creating your product by delivering the value manually. Imagine you are trying to build an online platform that connects French language tutors to students. Rather than building the software straight away, you can start recruiting tutors and connecting them with students manually. The goal with a concierge MVP is not efficiency but learning.

Working directly with your customers allows you to learn what they need, what challenges they face and which features to put in your digital solution. As you increase your customer numbers, you can move to using email to enable the connections. Over time you can start creating your digital product based on the learnings you have derived from using the concierge process.

We once worked with students in an innovation programme at the University of Kent (UK) who wanted to create a platform that connected students living in halls of residence to local mums who could make home cooked meals for them. They called

> ### 🔍 Useful tip
>
> The concierge MVP can be used to test specific assumptions. However, it can also be used to discover ideas for a solution if we don't yet have some.

their product Rent-A-Mum. But before building the online platform, they physically recruited some mums at a local nursery school who they paid to prepare the meals. They then sold the meals to some students on campus. Each transaction earned them very little, but they were able to test key assumptions about the product.

4.6 Don't look behind the curtains

Another method for testing our solution hypotheses when creating digital products is the Mechanical Turk or Wizard of Oz. This is similar to the concierge process, but there is one key difference. With the Wizard of Oz method, the customers get a digital experience via some interface. They have no idea that behind the scenes this value is being created manually. They just assume that there is a digital workflow underpinning the whole product.

Imagine that you are trying to create a product that improves the clarity of blurry photographs. Your plan is to have people upload their pictures and then your online software will automatically improve clarity. One way to start is to have people upload their pictures via a digital interface and you can have a team of people working in the background to quickly improve clarity and upload the pictures back to the customers. While the customer is having a digital experience, the value is being created manually.

The Wizard of Oz can be used to test several assumptions. For example, we can test whether our value proposition resonates with customers. But more importantly, we can test for the challenges we are likely to face when trying to deliver value to customers. What types of requests do they make? What types of problems do they present us with? How quickly do they need the service delivered? Are they satisfied with the solution or service we are providing?

All these lessons learned can be used to inform the features of our solution. Rather than build your entire product at once, you can then systematically automate certain features of the product. You can start with the features that promise the most value delivery. You can also take the opportunity provided by this process to test other aspects of your business model. For example, you can start to test pricing and channels for acquiring customers.

The founders of Zappos[28] ran a great Wizard of Oz experiment. Before creating the entire ecosystem for the business, they made a deal with a local shoe shop. The shop allowed them to take pictures of their shoes and post them on the website. Whenever a customer ordered, they would physically go to the store and buy the shoes, walk them to the post-office and ship them. Using this method, they were able to test several hypotheses about their business.

☐ What types of shoes are ordered and how difficult and expensive is it to ship shoes to customers?

☐ How often do shoes get returned and how difficult is it to manage shoe returns?

☐ Which channels drive the most customers to the website and which value propositions work best?

As they learned more, they were able to systematically design their business until it was automated and self-sustaining.

4.7 Co-Creation

Co-creation is about involving people from outside your company in the product design process.[29] These can be customers, inventors, suppliers, scientists or the general population. Collaborating with outsiders can be a powerful way to increase innovation capacity within an organisation. You are also more likely to get good ideas with a greater chance of success, especially if you collaborate with the customers who have the needs or problems you are trying to solve.

---

[28] Ries, E. (2011). *The Lean Startup.* New York: Crown Books.

[29] Benson, S. (2013). *Co-creation 101: How to use the crowd as an innovation partner to add value to your brand.* Vision Critical. Available at https://www.visioncritical.com/cocreation-101/

Examples of companies using co-creation

☐ DeWalt, the leading manufacturer of high-quality power tools, has an invention submission process where professional tradesmen and customers submit their ideas for new tools and product lines. DeWalt's process involves a review of all submissions based on uniqueness, commercial viability and fit within Dewalt.[30]

☐ Lego has an online community where customers can submit designs for new sets. Other community member can view and vote on the designs submitted by other customers. If a project gets 10,000 votes, Lego puts it through its review process and picks a winner whose design is manufactured and sold worldwide. The designer earns a percentage of the royalties and is recognised on all packaging and marketing materials.[31]

☐ DHL, the world's largest logistics company also uses co-creation to improve its supply chains and logistics. The company has hosted over 6000 hands-on co-creation workshops in Germany and Singapore with the customers and suppliers. These efforts have resulted in customer satisfaction scores rising to more than 80 per cent and delivery performance increasing to 97 per cent.[32]

☐ At Pearson, the global FTSE 100 education company, they have been working with young learners to co-design education products. In Pearson's Kids CoLab, young learners are treated as design partners. This means that their job is not to just give feedback on products designed by adult, but to be involved in a hands-on way by proposing design solution themselves.[33]

It is important to have a good process in place if you are going to be collaborating with outsiders. You need a process for people to submit ideas; either online or through workshops. You also need a process for reviewing ideas and making decisions. This is where the Lean PLC can be helpful in terms of setting the right criteria. There is also an

---

[30] See Dewalt Idea Submission Document. Available at https://www.dewalt.com/en-us/~/media/dewalt/files/company-info-documents/idea-submission-brochure-dewalt.pdf

[31] See Lego Idea. Available at https://ideas.lego.com/dashboard

[32] Crandell, C. (2016). Customer co-creation is the secret sauce. *Forbes Magazine.* Available at https://www.forbes.com/sites/christinecrandell/2016/06/10/customer_cocreation_secret_sauce/#3631ee8f5b6d

[33] Peterson-DeLuca, A. (2015). Putting learning at the center: Kids Colab Visits ISTE 2015. *Pearson Blog.* Available at http://www.pearsoned.com/education-blog/putting-learning-at-the-center-kids-colab-visits-iste-2015/

important role for enabling functions such as the legal department to play in negotiating deals for rights to customer ideas or for HR in terms of integrating new startup teams into your organisation.

4.8 Solution interview

During Explore, we conducted customer interviews to learn about customer needs. The goal of a solution interview is to learn whether we are thinking of the right product ideas to solve customer problems. Solution interviews can be done once we have started creating the minimum viable version of our solution. A solution interview is not a pitching exercise. It is about learning what is needed to build a great product for our customers.

☐ *Set context*: Thank the customer for taking part and then inform them of the goals of the interview. Recap your understanding of the problems you are trying to solve and check if these resonate with them. Inform the customer that you are working on a solution to solve these problems and you are looking to get feedback on whether you are heading in the right direction.

Q Tips

☐ Remember to go into each interview with orientation towards learning.

☐ Try to avoid interviewing multiple people at the same time.

☐ Do not ask customers for what features to put in your product - customers own the problems and innovators own solution design.

☐ Use calls-to-action to see if customers will perform the behaviours you expect.

☐ Use the opportunity to test other aspects of your business model such as pricing and finding the real decision makers/economic buyers.

☐ *Test solution*: Show customers the solution and allow them to use it. You can also demo the solution by showing how each of your features solves their problems. While you demo, ask customers for feedback. What resonated with them? What did they not like? What did they find difficult to understand.

☐ *Wrap up*: Thank the customer again for taking part. Ask for permission to contact the customer in the future. But most importantly, ask for referrals to other customers in their network who might have similar problems.

4.9 Presales

The ultimate measure of whether our product has a chance of success in the market is whether customers are willing to pay for it. This is an essential part of our business model. Most teams wait until they have finished building their product before they start selling it. But this does not have to be the case. You can start to testing your pricing model early by running a presales campaign. This when you sell your product to customers before it is ready.

How to do it

☐ *Crowdfunding:* This is a form of presales. This works particularly well with consumer products. This method has been used to raise money to build the product. Raising these funds is a great way to get a consumer vote of confidence. There are several crowdfunding platforms available. For example, Publishizer allows authors to crowdfund support for their book before they have finished writing it.

☐ *Landing page*: One use of the landing page can be to test pricing. We once worked with a team that wanted to test whether people would sign up for language classes. They used Eventbrite to sell seats in the class before they had even developed their materials. This helped them gauge authentic interest in their product, as well as the right price point to sell it.

> ☐ *Letters of intent and pre-ordering*: If you are working on a product that will be sold business-to-business (B2B), then getting signed letters of intent to buy from customers can be a useful signal. Even better is to get a business to pay you a fully guaranteed deposit for the product (e.g. 25–50 per cent deposit).
>
> There is nothing that beats real money in the bank as a measure of customer intentions to buy. It is possible to measure this be asking the right questions. However, behavioural measures of buying are more reliable than verbal promises about the intention to buy. It is important to note that presales are not the same as launching a full-scale marketing campaign to the wider public. Such experiments have to be systematically run using early adopters or a small segment of the main target market.

4.10 A/B Testing

One key criterion for having actionable metrics is comparability. We need to be sure that the incremental changes we are making to our product are moving the needle in the right direction. This can only be done by comparing customer responses to a new version of our product, against the previous version. It is also the case that there may be one or two different options for improving our product; and we are not sure which option works best. A/B testing can help us resolve this by testing how customers respond to certain features (or price points) over others.

> ### How to do it
>
> The name A/B testing describes exactly what the process is. You have two or more versions of a product and then you run an experiment that allows you to compare them on the basis of a predefined metric (e.g. conversions rates, bounce rates, signups, downloads and purchasing):
>
> ☐ *Have a clear goal*: A/B testing should not be done without a clear aim or learning goal. Every experiment should be focused on testing specific aspects of our product and how that impacts our target outcome.

☐ *Focus on the biggest pain points*: In the early days, our focus should be on removing the barriers to making our business model successful. The colour of a button may not be as important as whether or not our value proposition is leading customers to sign up and pay for our product. So, it's important that our A/B tests focus on meaningful outcomes; revenue matters more than conversion rates and profits matter even more than revenue.

☐ *Random assignment of customers*: We need to use tools that enable us to create different versions of our product and present them to customers. Whatever tools we choose should also be able to randomly assign customers to the various versions of our product and measure their responses. It is key to remember that we have to test the different versions of our product at the same time. We should not do our testing linearly (e.g. start with one version and then test the next one).

☐ *Control the conditions*: When running an A/B test it is important to have some control over the variables that are being tested. Even as we iterate our product, we should be doing this in a systematic way. The focus should be on one or two factors that we manipulate and change for each specific A/B test. If we change too many things across our versions, we will not know which of those changes have resulted in the outcome we are observing.

☐ *Old visitors versus new visitors*: As we set up our A/B test, we should be conscious of whether the customers we are running the test on have seen our product before. One thing we don't want to do is surprise our current customers by exposing them to ever-changing price offerings! In this regard, we should consider testing any new version of our product on new visitors only. It is still possible to run experiments on current customers when we plan to launch new features. However, once they have seen one version of the A/B test, they should not then be exposed to a different version, if they should return to our website.

☐ *Patience is a virtue*: As we run the A/B test we should have the patience to let the data collection play out. Early on, it may seem quite clear that we have a winner. But due to random error, such early signals should not be used as basis to stop the A/B test. We should set sample size target that allow us to have statistical confidence. There are online tools that can be used to determine the right sample size. A rule of thumb that we can use is at least 30 people per condition. This is based on the statistical rule that you need at least 30 people in a sample before you can reasonably expect your analysis to be based upon a normal distribution. This allows you to be able to make statements about statistical significance.

☐ *Multivariate testing is an option*: Just because it's called an A/B test doesn't mean that you only test two versions of your product. As long as you do it in a controlled way, you can test three or four versions. You can also test more than one variable. For example, you can test the combination of two different value proposition statements and two different price points, to see which combination works best. The point is to be systematic and keep everything else exactly the same, outside of the variables we are testing.

☐ *Make decisions based on data*: A/B tests can produce counterintuitive and unexpected results. But even if we don't like the result we get, we should not use our own biases to overrule the findings! This negates the whole purpose of running experiments and can quickly take us down the wrong path. It is also possible that our test can produce inconclusive results. Again, when this happens we should not use our own biases to make a call. It is better to iterate on the experiment and run it again.

Always remember that metrics are people too. A/B tests are great for providing quantitative data about what people will do. However, they are not great for finding out the underlying reasons why people are making those choices. Therefore, it is still important to talk to some customers after the A/B testing is complete. Understanding the underlying reasons for people's behaviour will also help us with designing better experiments as we iterate our way to success.

4.11 Usability testing

As our product begins to take shape, it is important for us to start thinking about the user experience. The best product experiences are intuitive in the sense that they do not require customers to do too much thinking. There is really no other way to figure this out, except to give customers early access to our product, watch them use it and see which aspects they find difficult. These learnings can then be used to improve the product. This is usability testing; and it is an important part of product design process.

How to do it

☐ *Start testing early*: There is never a wrong time to run a usability test. As soon as you have something to put in front of customers, do it. It is much better to pick up problems early, fix them and get on the right path; rather than do it later when the product might be more complex.

☐ *The right people*: Do not run a usability test on members of your own team. You have to identify and recruit real customers in the real world. The use of early adopters is highly recommended in the early stages of development, with main market customers being more helpful later.

☐ *Be prepared:* It is important to have clear learning goals and objectives around usability testing. With clear learning goals, you can write a script that outlines what is going to happen during the test. During the session itself, you don't have to read from the script. But it is important to be prepared.

☐ *Participant comfort*: We are testing our product not the customers. So, this is something we need to make clear at the beginning. We must encourage customers to be honest and open and tell them that there are no wrong or right answers. During the testing itself our tone must remain positive and encouraging. We must not do or say anything that may cause our participants to become uncomfortable.

☐ *Concrete scenarios*: Rather than asking participants to simply 'play' with our products we should ask them to complete specific tasks related to our assumptions. For example, customers can use our product to perform task such as registering for an event, paying bills or uploading a document. The point to test how well our product helps customers complete their jobs.

☐ *Listen and learn*: During a usability test, the customer has to talk more than we do. We want to learn and so we must encourage the customers to verbalise what they are thinking as they are using our product. We should not be influencing or helping customers with the task at hand. We must remember that we will not be there when our product is in the wild.

☐ *Team observation*: It is important to get members of our team to observe the usability testing session. Teams often argue about the features of the product; especially what will work. However, watching real customers using the product can sometimes resolve some of those arguments.

☐ *Record session*: With customer consent, it can be useful to record the session. This means that the host can concentrate on interacting with customer, rather than taking notes. Recorded sessions also make it easier to perform analyses, compared to trying to recall details from memory, as this can be influenced by our biases.

There is often a feeling that usability testing can be a long, complex and expensive process. However, testing can be fast and inexpensive. First, you don't need to hire an external consultant to run your usability testing for you. With the right training, internal teams can run their own tests. Second, you don't to have large sample. According to Jakob Nielsen, you can identify the majority of the usability problems in your product with just five users.[34] The point is to test early and often.

---

[34] Nielsen, J. (2012). *How many test users in a usability study?* Nielsen Norman Group. Available at https://www.nngroup.com/articles/how-many-test-users/

4.12 The minimum – the sellable

Early in this chapter, we described minimum viable products as instruments for running experiments. Even after talking to customers and understanding their needs, the exact solution that will deliver value takes time to emerge. This is why we use minimum viable versions of our product to begin to test our early assumptions about the solution. The goal of this process, and all the techniques we have described so far, is for that clarity around the right solution to emerge.

When we begin working on the early versions of our minimum viable solution, our goal is not to get traction. Our goal is to learn and systematically iterate towards those features that will provide traction. We use experiments to figure out the value proposition that will resonate, the features that will deliver value, how much people may be willing to pay and the right user experience for our customers. Over time, we want to create a version of our product that customers love. But most importantly, we want to create a product that customers will pay for.

This minimum sellable product must contain the necessary features and user experience to tip customers over the edge and make them reach for their wallet. And so over time we need to focus on the additional functionality that will make our product saleable. What are the table stakes for our customer, beyond our core value proposition? This can include things such as security, ability to integrate with currently used solutions, ease of adoption and low switching costs.

Figuring this out needs to be done systematically using the experimental process we have described in this book. However, is it also important that we avoid creating bloated products! Don't do too much to get the sale, do just enough. Remember that we want to start generating revenue and get to sustainable traction. We will have plenty of time to add bells and whistles during the Grow and Sustain stages. Making these decisions is not an exact science. It is ultimately a judgement call.

It is important to emphasise at this point that when we refer to minimum sellable product, we are not talking about how to trick our customers into buying. Our focus is still on delivering value. We want our product to be sellable because we have made something people want. So as much as we will measure whether we are getting sales, we should also keep a close eye on metrics around customer engagement, retention and user experience.

4.13 The Funnel

As we develop our minimum sellable product, we have to think more deeply about how we are going to sell it. One lens that we can use is the funnel through which our customers hear about our product and move towards purchasing it. There are several examples of such funnels, but the most popular one within Lean Startup is Dave McClure's Pirate Metrics.[35] This funnel goes through five key steps.

☐ *Acquisition*: How are customers hearing about us? Are we acquiring new customers in sufficient numbers to make our product viable?

☐ *Activation*: When we have acquired the customers, are we able to get them to engage with our product (e.g. try it out, sign-up, use it)? When we activate them, are they having a good experience with our product?

☐ *Retention*: This is about bringing people back to our product. Are we retaining customers and getting repeat business from them?

☐ *Revenue:* Are customers willing to pay for our product? If so, how much are they will to pay and is this enough revenue for profitability?

☐ *Referral:* Are customers happy to tell other people about our product?

Figure 4.1

How do customers find you?	**Acquisition**
Do they have a great first experience?	**Activation**
Do customers come back?	**Retention**
Are customers willing to pay?	**Revenue**
Do they tell others about the product?	**Referral**

---

[35] McClure, D. (2010). MoneyBall for startups: invest before product/market fit, double-down after. *Master of 500 Hats.* Available at http://500hats.typepad.com/

How to do it

These are known as Pirate Metrics because if you put together the first letters of each stage its spells AARRR, like that favourite pirate phrase! The great thing about this funnel is that it forces innovation teams to focus on only five business relevant metrics that matter, rather than have a dashboard with hundreds of numbers that do not help us make decisions. As we run each experiment, we can be tracking the impact our work on these five stages to see if we are making progress.

Of course we can't focus on all five metrics at once. This is also not a linear process where you begin with acquisition then work your way down to referrals. Rather, we have to focus on the right things at the right time. For example, before we know that we have created something of value that people will like (i.e. activation) and want to keep using (i.e. retention), then it makes no senses to focus on improving our acquisition process. If we do that, this will result in plenty of unhappy people who could be lost to us forever.

What we must do instead, is drive just enough people to our product to begin learning whether we can activate and retain them (and maybe even get them to buy!). After we have run a series of experiments to resolve this, we can then turn our attention to ramping up our acquisition engine. This is also not a one-time thing, we must keep iteratively improving on these five key steps, even as we enter the Grow and Sustain stages of the lifecycle.

4.14 Cohort analysis

One of the hardest things for innovation teams to do is track whether they are really making progress. Teams will be making improvements to their product and running experiments to test their hypotheses. Using Pirate Metrics and A/B testing as we described earlier is one way to measure progress. However, as our various experiments pile up, how do we know we are truly making progress and improving our key business metrics over time.

If we use aggregated data, we can be fooled into thinking that we are doing well. We start with 100 customers and if next month we get 45 new customers and lose only 15, we will have 130 customers. This number can keep going up and to the right because it is aggregated. However, if we are working on product and 'improving' it each month, then our new (and old) customers will be getting exposed to different versions of our product.

So, the real question is not whether our customer numbers are going up in an aggregated sense, it is whether the changes we are making are improving our customer activation and retention rates. In other words, are we improving at an improving rate. If we were retaining 70 per cent of our new customers three months ago, have the improvements we have made to the product improved this retention rate to 75 per cent or more? If we were getting 25 per cent of our customers to part with their money, have we improved this conversion rate to 30 per cent or more?

In other words, are we really getting better? The only way to answer that question is to use cohort analysis. A cohort is a group of customers that share the same characteristics or receive the same treatment. For our product, a cohort can be group of customers who have been exposed to the same version before we make changes and improvement. Cohorts can be tracked weekly or monthly, depending on the frequency of our experiments.

Figure 4.2

	Month 1	Month 2	Month 3	Month 4	Month 5	Month 6	Month 7
(Joined in) Week 1	100%	10%	9%	9%	7%	7%	7%
Week 2	100%	12%	10%	10%	8%	7%	?
Week 3	100%	16%	14%	13%	12%	?	
Week 4	100%	17%	15%	14%	?		
Week 5	100%	20%	19%	?			
Week 6	100%	22%	?				
...	...	...	...	...	...	...	...

The power of cohort analysis is that it shows more clearly whether our work to improve the product and business model is having the desired impact. For example, last month's cohort can show us that we have a 25 per cent conversion rate at the revenue stage of the funnel. If we do some work to improve our website, then the impact of this work will be most clearly visible in the next cohort of customers (e.g. conversions improve to 28 per cent). This combination of Pirate Metrics and Cohort Analysis provides the best kind of data for making decisions. We can really see how our funnel is improving from cohort to cohort over time.

4.15 The business model revisited

When we began the Validate stage of the Lean PLC we had several business model assumptions to address. All the various tools and methods that we described in this chapter have been aimed at transforming our key assumptions into knowledge. For example, we used the co-creation, concierge and Wizard of Oz methods to discover the right solution for our customers. We also used our pre-sales and A/B testing to test our value proposition, revenue models and channels to customers.

Our toolbox also helped us to find the key partners, key resources and key activities we need to create our solution. This would have happened as a by-product of our efforts to create the right solution. Our funnel metrics showed us the cost of customer acquisition, conversion rates and retention rates. At the end of this process we should have some sense of how much it will cost to create the solution and what customers will pay for it.

We are now faced with a key decision. Are we ready to scale? How confident do we feel that we have more knowledge than assumptions in terms of our business model. Do we think we can get to sustainable profitability? The answers to these questions will never be perfect or absolute. There will always be some element of taking a leap of faith. However, unlike the traditional business plan, at least our decisions are based on some evidence and validated learning.

We should also remember that having customers and a solution they like is not the same as having market. As it becomes clearer who our customers are likely to be, we should use that information to revise our market size estimates. This will inform our decision to stop, pivot or move to the Grow stage. If we feel we are ready to scale, then it is time to prepare for Grow!

4.16 Preparing for Grow

Wow. What a journey! After all the long hours working on your idea, testing assumptions, pivots and iterations, you are now ready to scale. Unfortunately, your work is not done yet. It is just beginning. A lot of teams fail, even after they have created something customers want. This because we are not only looking for a profitable business model, we are also looking for a scalable one. Resolving this question is what you are going to have to do during Grow.

This is the point at which you present to the product council something that is similar to the typical business case. Your request for resources to scale should be based on the lessons learned during Validate and your hypotheses about the level of growth you can achieve over the next 3–5 years. Your submission should include:

- [ ] an update of the work we have done so far and how the evidence supports your decision to scale

- [ ] a detailed description of the product, how it creates value for customers and the strategic goals of our company

- [ ] an update on the target customers for the product and expected size of the market opportunity

- [ ] our hypotheses about how much we will growth customer numbers, revenues and profits over the next 3–5 years

- [ ] a high-level roadmap of any new features you plan to add to your product and how this will help you meet your growth targets.

- [ ] a full P&L spreadsheet that will serve as the initial benchmark for future analysis and decision making

- [ ] the resources and funding we think we will need to complete to take the product to scale in the first year.

SUBMISSION TEMPLATE

Validate to Grow SUBMISSION

Idea ownership

Investment board	
Business sponsor	
Product owner	

Idea overview

Product name	
Idea description	
Strategic fit	

Target customers and jobs to be done

Describe your potential customer segments and their needs or jobs to be done.

Customer segments	Jobs to be done

Validate lessons learned

Provide an overview of the assumptions you tested in Validate and the lessons learned. Add more customer assumptions as necessary.

We believe that:	
To verify that we:	

And we measure:	
We learned that:[36]	

Unexpected learnings

Provide a summary of the unexpected learnings you got during Validate.

Evidence of problem-solution fit

Provide any evidence that the solution you have created meets customers' need and that customers are happy with it (e.g. positive reviews, usage metrics, retention metrics and positive impact).

Evidence of problem-solution fit

Provide any evidence that you have product-market fit and that your business model has potential to be profitable (e.g. customer commitments, early revenue or breakeven).

---

[36] Osterwalder, A., Pigneur, Y., Bernarda, G. and Smith, A. (2015). *Value proposition design*. New York: John Wiley.

Market opportunity

Please update your learnings about the market opportunity below:

Total addressable market (TAM) in dollars ($)	Serviceable addressable market (SAM) as % of TAM	Serviceable obtainable market (SOM) as % of SAM

Market description

How much did you spend in explore

Provide a breakdown of your spend during Validate.

What are your plans for Grow?

Please provide a summary of the activity you plan to do next.

Assumptions to test during validate

Provide an overview of the assumptions about the business model you plan to test during Validate, how you will test them and your success criteria.

We believe that:	
To verify that we:	
And we will measure:	
We will know we are right if:	

Financial projections

Since you are asking to move to Grow, please complete the financial data below.

Expected revenue over 3 years	Expected costs over 3 years	Profit margins over 3 years (%)

Resources and funding requested

To complete the Validate stage, we are asking for (e.g. **dollars, time, people**):

* Please note that you also have the option to remain in Validate, go back to Explore or stop the project entirely. If either of these options are what you are choosing, then you need to adapt this template so that you can update the product council on work done so far, key lessons learned, what you plan to do next and the resources you need

? An interview with:

Teodora Berkova, Director of Social Innovation, Pearson

Pearson is on the FTSE100 and is the world's largest education company. In 2016, Pearson launched the Tomorrow's Market Incubator with a goal of reaching underserved low-income and emerging middle-class consumers across the globe with great education products. We spoke with Teodora Berkova who is the director of the programme on how they used Lean Thinking to manage the process.

Please tell us a little bit about the social innovation project you are running within Pearson?

The need for learning is relevant to everyone, no matter where they live or how much they earn. Today, however, Pearson's products and services do not reach many of the four billion low-income and emerging middle-class consumers across the globe. To bridge this gap, Pearson aims to innovate new business models that profitably deliver affordable, high-quality education products and services built on Pearson's world-class educational technologies. Developing products and services for these segments is different from traditional product innovation approaches. While keeping to the lean and agile methodology, within Pearson they often require full-scale business-model innovation, new product formats and technology platforms, unconventional distribution channels and revenue models.

The Tomorrow's Markets Incubator (TMI) is an innovation platform designed to support the development of long-term business opportunities that require this type of extensive innovation. The TMI supports employee-driven, early-stage ideas from across the company. We partnered with TIL Ventures (a firm which helps corporates explore and create new markets), to help us setup and design the TMI. Ultimately, our goal is to provide a platform for the Pearson people who are on the ground and who have a deep understanding of unmet consumer needs, to propose relevant product and service ideas and to empower them to develop these ideas.

How did you source ideas for the project? How did you make sure that most people within the company could participate?

The Incubator was open to any Pearson employee, based anywhere in the world to participate. We launched in April 2016 via an internal communications campaign with global reach, including a webinar open to all interested employees and a website where employees could learn more and submit their ideas. The webinar explained the purpose of the Incubator and the one-page initial idea application.

How did you select the ideas that eventually got investment? How did you apply the Lean Product Lifecycle to your decision making?

The investment committee was looking for early-stage ideas and we based the phases of the Incubator – Idea, Explore, Validate, Grow – on Pearson's Product Lifecycle. We adapted some of the criteria and application materials to address (1) the unique needs of low-income markets with limited resources and infrastructure and (2) the market creation challenge where a given education product/service functionality might be entirely new for the respective consumer. The vetting process also took into account the potential for scale. This is especially important given the higher risks and longer return on investment, which require that the opportunity is financially compelling for the company.

The criteria for investment were as follows:

1. Mid- to long-term profitable revenue potential
2. Focus on low-income learners and consumers
3. Alignment with Market and Business Unit strategy
4. Potential for scale and feasibility within Pearson

What were the sizes of the initial investments and what were the expected milestones that teams were supposed to reach?

Each team received between £30,000 and £40,000 to support research needed to build out their ideas. Each team also worked closely with the Tomorrow's Markets Incubator team and other coaches. During the Explore stage, teams were expected to conduct research with learners and customers and identify: a lever customer and learner segment, their pain points and daily routines; potential at-scale customer base;

product/service concept; proposed business model; whole cost structure so that we could estimate the full cost of running the venture; efficacy framework outlining impact on learning and other social impact indicators.

If your experience, what did most teams struggle with when testing their ideas? How did the best teams handle these challenges?

Teams struggled with knowing how to design, set-up and conduct research with potential customers and learners in the early stages. Given that various participants had very different backgrounds, in addition to providing help with early-stage deep consumer research, where needed we also provided coaching on business model development, sales, marketing, product development, financial modelling, social impact measurement, leadership and management skills.

Out of the initial cohort that got funding, how many teams got follow-on funding to keep working on their products?

Out of the initial 17 teams which went through the Explore stage, our panel selected four teams to receive up to £250,000 with which to rigorously demonstrate proof of concept of their venture during the next phase.

How did you ensure that teams had support from Pearson business units for the product ideas they were working on?

We coached teams on engaging key internal stakeholders from the very beginning and getting their input and buy-in even before needing explicit support from them.

As you prepare for your next cohort, what lessons learned are you going to take forward? What will you do differently or better?

We will have a team of experienced coaches who can dedicate significant time to working closely with the teams in their respective markets from the get go – previously it took a while to get this set up and ready to go. We're also going to be experimenting with different ways to help teams generate ideas ahead of submitting their idea forms. This may entail setting up collaboration labs with other companies, or issuing specific education innovation challenges.

Part 2

5

Grow

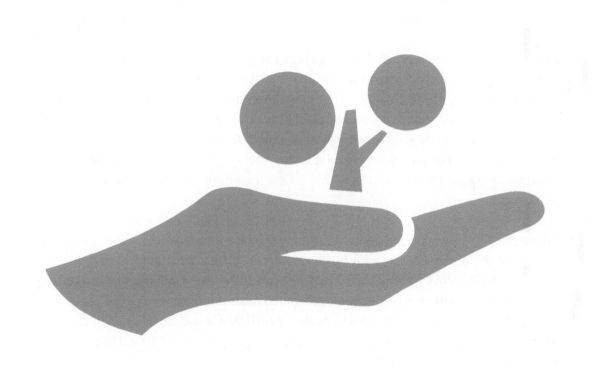

5.1 Welcome to Grow

Welcome to the right side of the Lean Product Lifecycle (Lean PLC)! As you went through Idea, Explore and Validate, you ploughed through the paths of uncertainty using experimentation and iterative product development methods. Emerging from that journey, you now have some clarity on the right solution for your customers and the right business model to deliver that value. This is the transition point from the left side of the PLC to the right side. It is now time to double down on your validated business model. But if you think that this is all about execution; well think again!

Grow will be a critical and challenging stage for your product. It is not as simple as pouring fuel in your rocket and watching it fly. This is the lifecycle stage where many of the technology game changers that have been documented in the media such as Uber, Airbnb and Dropbox really knocked it out of the park.[37] But this is also the stage at which a lot of great products have failed. Grow requires relentless focus, energy and commitment over the long term. And even as you scale, you will need to continuously apply lean thinking within an environment which is constantly evolving.

To succeed in Grow, you will need to navigate the operational and cultural challenges of discovering and executing your business model at the same time. You will need to continuously improve operations so that you can scale fast. It will require you to keep pace with the progression of technology against a continuously evolving and relentlessly changing business environment. And if you stand still for too long, someone will be ready and waiting to take your customers. The Grow stage can be split into the following key phases.

1. *Tuning the engine*: This phase focuses on finding the right growth engine and testing our growth hypotheses. We will have learned some key lessons around potential growth during Validate, but now we want to refine our learnings so we can fully execute on our growth strategy.

---

[37] Bulygo, Z. (2012). *The 7 ways Dropbox hacked growth to become a $4 billion company.* Kissmetrics Blog. Available at https://blog.kissmetrics.com/dropbox-hacked-growth/

2. *Accelerating growth*: As our knowledge and confidence around our strategy increases, we can then turn our attention to accelerating growth. We can deploy various growth tactics to rapidly increase customer numbers, revenues and profits. The goal here is to not only grow these key numbers but also improve our growth rates.

3. *Optimising the engine*: When we get to full velocity, our product will be growing fast. This does not mean that we rest on our laurels. We will continue to improve the product and optimise our growth engines. Such continuous improvement is a key part of lean thinking; and we will use it to maintain our growth rates.

4. *Preparing for sustain*: Every product eventually matures in the market. Growth rates will slow down as technologies and customer preferences shift. This does not mean we kill our cash cow. Rather, it means we have to change how we manage it. Our investments in growth can now be decreased and we can turn our attention to optimising and reducing costs.

5.2 Premature optimisation

So far in this book, we have been cautioning you against premature scaling. If you have followed our advice, you have worked hard to systematically uncover a great business model for your product. While this may be exciting, we have some more words of caution for you: *avoid premature optimisation.* In most companies, when a team with great product enters the Grow stage, they come under immediate pressure from leadership to rapidly grow the product. This can lead to early standardisation and counterproductive marketing expenses before the team has found the right growth engine. In other words, avoid those Super Bowl ads.

Before scaling rapidly, what we want to do first is find sustainable growth. According to Eric Ries, a product has sustainable growth when new customers come from the actions of current customers.[38] This is different from one-off marketing activities that may drive an initial bump in customer numbers and then fade over time. There are various ways to achieve sustainable growth and we will discuss these in more detail later. At the moment, we want to emphasise that before finding sustainable growth, any actions towards ramping up scale are bound to fail.

---

[38] Ries, E. (2011). *The lean startup*. New York: Crown Business.

To find sustainable growth, we have to apply the same build–measure–learn process that we used during the earlier stages of the Lean PLC. We have to identify clear hypotheses about the level of growth we want to achieve and how we think we might achieve that. We then have to test these assumptions by deliberately measuring the impact of each growth tactic on the bottom line. Our goal is to find a repeatable process for driving customers to our product, converting them into paying customer and retaining them over the long term. Only after we have found this repeatable process are we then ready to scale.

As such, we never stop using lean thinking across the entire Lean PLC. As we grow our product we have to manage a continuous cycle of learning and development.

1. We will need to ensure that our teams are balanced and resourced so that they can simultaneously manage searching and executing on your business model.
2. We will need to decide on a range of key metrics that we will track to manage growth and scaling.
3. We will benefit from introducing a cadence for growth reviews during which we examine whether our growth tactics achieve our hypothesised results.
4. We will need to continuously advance our practices so that we can reach a state of continuous delivery.

5.3 Beyond the early adopter

As can be seen from Figure 3.2, on the left side of the Lean PLC you did most of your work with early adopters. However, in order to scale, you have to start thinking about mainstream customers. Unlike early adopters, mainstream customers are more demanding in terms of product quality and reliable service. The channels and marketing messages to reach mainstream customers will also be a little different to those you used to reach early adopters; and as your customer numbers increase you cannot retain the intimacy you once had with customers.

As such you need to adopt a different mindset. All of your learnings from Validate will need to be reconfirmed as you take the product to scale. This includes the technology stack, levels of service provision and business model which will need to be continuously refined as you learn. When maturing operations for scale, your team's culture will be challenged and will be expected to evolve. In most cases each challenge may slightly change the culture and you will need to be mindful of the direction it takes. These changes are needed to grow, but also risk negatively impacting growth if not managed well.

To go beyond innovators and early adopters there are several things that you now need to consider. Table 5.1 lists things which your early majority customers and beyond will expect over your early adopters which you'll need to prepare for.

Table 5.1   Early adopters versus early majority

	Early adopters	Early majority
	Takes higher risks	*Takes lower risks*
1	Open and willing to try the product and excited to be involved with a new innovation. Willing to apply the product to many applications.	Focused around what the product can do for their specific job to be done and less excited about a new technology novelty. Will need a more obvious application to the job to be done.
2	Willing to pay a higher price than adjacent competitors, to be the first to access a new technology.	More price-aware and will be actively comparing a broader range of competitive solutions.
3	Open to trying the product with limited or no marketed company presence.	More assurance of the company behind the product and brand. Likely to want to see marketing presence and market domain association.
4	Willing to tolerate some failures and willing to spend time to work around minor problems and configuration issues.	Expects a higher level of product quality with provisioned support and services. Expects the product to not fail when used.
5	An adoption leader is willing to make choices which are unpopular and new as an individual.	Is part of a crowd and will follow the trends. Considered a follower.

5.4 Engines of growth

In order to grow effectively, companies need to have a shared language. As part of the lexicon, innovation teams need to understand the growth engines for their product. An engine of growth is our system for sustainable traction. It is how we ensure that the actions of current customers are consistently driving new customers and revenues to our product. In other words, engines of growth are a repeatable system for growing customer numbers, customer retention, engagement, revenues and profits. There are three main engines of growth that teams can use.

- *The paid engine* relies on paid advertising, marketing or a sales force to grow your product. With this engine, the key behaviour from current customers that drives growth is the amount of money they spend when they purchase our product. Since we are paying to acquire customers through advertising, marketing or a sales force, there are costs associated with every new customer we get. To achieve sustainable growth, the cost of customer acquisition must be less than the money those customers spend when they purchase our product. This then allows us to reinvest the money to acquire more customers creating a virtuous cycle of sustainable growth.

- *The sticky engine* relies on customer retention. The challenge here is how we get customers to stay with our product and potentially increase our share of their wallet. With this engine, the key behaviour from current customers that drives growth is whether they keep coming back to use our product. For sustainable growth, our retention rate has to be higher than our churn rate. If the rate at which we acquire and keep new customers is greater than the rate at which we lose customers, then sustainable growth will be achieved.

- *The viral engine* relies on word of mouth and usage by your current customers to drive growth. With this engine, the key behaviour from current customers that drives growth is the extent to which customers are willing to actively market our product for us by making referrals. The viral engine can also work when current customers inadvertently market our product through their active use of it. The success of our viral engine is indicated by our viral coefficient. In a good viral engine, a current customer brings with them on average more than one other customer (e.g. 2 or 3). This means that any viral coefficient above 1 indicates a good engine with sustainable growth.

It is possible for a product to have more than one growth engine. For example, a viral engine works really well if you have also got a sticky product with good retention rates. The key is to identify your primary engine and use that to make decisions. Our goal is to ensure that our growth engine works the same way with mainstream customers as it did with early adopters. Getting your growth engine right and refining it regularly is really important. For example, Dropbox had to pivot away from their initial paid engine of growth to a viral engine of growth. This change increased their user base from 100,000 to 4 million in 18 months.[39]

It is also important to recognise that sustainable growth has to be based on sustainable profits. A great viral engine with high retention rates is a good indicator that customers love our product. While we can grow customer numbers with a good viral engine, we don't want to do this while losing money! It is true that some great companies have been built through an initial focus on growing customer numbers (e.g. Facebook). But this growth was not an end in itself. It was simply an initial step to grow customer numbers with a view to eventually monetising those customers and becoming profitable.

5.5 Tuning your engine

When you first enter Grow you are still learning and your product could still be changing in response to the market. It is possible to view the first stages of Grow as beta. The best way to avoid premature optimisation is to benchmark where you are at the moment. You may not yet have the large numbers of customers that indicate a growing product. What we need to know at the moment is the product's potential for growth and make decisions about what to do next.

In order to tune our engines of growth, we have to revisit our funnel and cohort analyses. As can be seen in Figure 4.1, we are looking to examine our per-customer metrics. What are our customer acquisition rates using our current marketing channels? What is our average cost of customer acquisition? How well are we

---

[39] Kissmetrics Blog (2016). *The 7 ways Dropbox hacked growth to become a $4 billion company*. Available at https://blog.kissmetrics.com/dropbox-hacked-growth/

converting or activating our customers? What are our current customer retention rates? How much revenue are we getting per customer? What is our customer lifetime value? How many referral customers are we getting? What is our viral coefficient?

If we benchmark these numbers, we will know what the current state of our product is. We can then begin to work towards growing the product while tracking any improvements in these numbers. The cadence at the which we track the numbers depends on the number of growth experiments we are running and the pace at which we acquire new cohorts of customers (e.g. weekly or monthly).

As noted in the previous chapter (Figure 4.2), we want to track our metrics on a cohort by cohort basis. We do not want to mix last month's customers with the new customers acquired this month. The reason for separating our cohorts is that we want to see if there are any real improvements on our per-customer metrics. Are we growing the product at an improving rate? This benchmark becomes the basis for decision making going forward. The success of every decision we make will be judged by our ability to move that needle.

Things to look out for in the first sub-phase of Grow:

1. The customer expectations of quality in terms of your product and supporting services have significantly increased and will continue to do so.

2. The engine of growth, customer behaviours and business model defined in Validate will be subject to change as you expand to the Early Majority and beyond.

3. The cost and effort to maintain and transfer knowledge across your team will increase as your team sizes grow.

4. Your customer support, services and sales teams need to be laser focused and coordinated constantly. This is especially required where many parts of your business will have direct contact with customers where others won't; increasing the risk of assumption and interpretation.

5. You will need to advance and refine technical operations, services and performance constantly for repeatability and scalability.

6. If you are part of a larger enterprise or have investors, you will need to actively obtain and maintain explicit senior support and buy in to maintain momentum on a regular basis.

5.6 Growth hypotheses

In most traditional business plans, you will find growth projections. These are usually described in 3–5 year periods. Within the Lean PLC, we prefer the team use growth hypotheses instead. This minor difference in wording represents a significant difference in orientation. While growth projections can result in an exclusive focus on execution, growth hypotheses orient teams toward learning and discovery as they scale.

It is important to note that our growth hypotheses are not pulled out of the air. They are based on what we learned during validate and what we need to learn going forward. Much of what you think you know is an assumption and not necessarily fact. At the same time, we need to have a vision about where we need to be for the product to be successful.

Figure 5.1

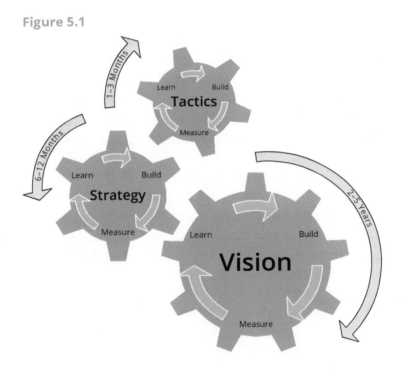

There are three main time layers within our hypotheses. The first layer is our 2–5 year vision of where we want the product to be. The second layer is our annual growth strategy that will help us achieve our long-term vision. Finally, the third layer is more tactical and refers to shorter time cycles such as monthly or quarterly growth numbers that are related to our annual growth hypotheses. The time to validate each of the layers varies and as such, each of these three layers have their own set of hypotheses which should have a direct connection to each other through leading indicators. If tactics need to pivot when reviewed at a quarterly point, this doesn't and shouldn't mean your vision needs to. If you fail to deliver tactical success on a continuous basis though, this might indicate a change in strategy is needed.

Our growth hypotheses have to be specific and actionable. As part of our 2–5 year vision we have to set clear goals of the customer numbers, market share, revenues and profits we want to achieve. Once we know our long-term goals, we can then break it down into our per-customer metrics or business relevant units of value (e.g. customer numbers or number of new subscriptions).

Imagine a team that wants to achieve $50 million in annual revenue by year three. This broad number can be broken down into the number of customers, sales, or subscriptions needed to achieve that goal. For example, the team using a subscription model can calculate that they need 200,000 customers paying $250 per year to reach their target of $50 million in annual revenue. This larger growth target, together with the current state of the business, can be used to set annual growth hypotheses (e.g. Year 1 = 50,000 customers at $150; Year 2 = 125,000 = customers at $200, Years 3 = 50,000 new customers at $100 CLV (customer lifetime value)).

At the tactical level the annual growth hypotheses can be then broken down further into monthly hypotheses. Now the team has clear benchmarks with which to monitor progress and learn early whether they are on track to reach their goal. As the team deploys its growth tactics, it can set clear per-customer target metrics such as conversion rates, retention rates and lifetime value. Each cycle of tactical growth experiments can then be tracked against these minimum success criteria.

🔍 Tip

Although metrics are needed to validate your hypotheses, it's recommended that tactical reviews incorporate subjective learnings. Asking and recording explicitly what a team has learnt at each review point, will help construct a narrative to be explored which will inform strategy. This practice also provides reassurance that the teams are actively learning and having the ability to apply new approaches which reduces waste and risk.

5.7 Growth tactics

Once we have outlined our growth benchmarks, we can then begin implementing and optimising our growth engines. On the basis of our funnel we can develop techniques to acquire, activate and get customers to pay for our product. There are also techniques that help us retain customers and convince them to make referrals on our behalf. As we deploy each tactic we should do it in a systematic way, while keeping an eye on our target metrics.

Patel and Taylor[40] differentiate between pull and push tactics for acquiring customers. *Pull tactics* involve subtle marketing techniques such as blogging, writing magazine articles and social media campaigns. The goal is to give customers a reason to come to our product through enticements and incentives. In contrast, push tactics are more aggressive and involve acquiring customer in more direct ways such as sales teams. Both pull and push tactics apply are part of the paid engine of growth. As such, when we use these tactics we have to ensure that our cost of customer acquisition is lower than the customer's lifetime value.

The viral engine of growth can also be used when the product becomes its own customer acquisition tool. Product-based tactics include customers inviting friends to our product, enabling customers to talk about our product on social networks and incentives such as giving customers some reward if they make referrals and get their friends to sign-up. It is important to note here that acquiring customers is not the end goal. We want to activate them to buy and use our product.

To activate customers and get them to buy our product, we must have clear *calls to action.* Once the customers are engaged it must be clear what is expected from them. This applies to both physical and digital products. Customers need to know how they can buy the product, sign-up or subscribe if necessary. For complex products, it is also important to have a good process for onboarding customers which reduces the customers' time to value. A good onboarding process can increase customer activation rates. Teams should provide support and guidelines to help customers use our

---

[40] Patel, N. and Taylor, B. (2017). *The definitive guide to growth hacking.* Quicksprout. Available at: https://www.quicksprout.com/the-definitive-guide-to-growth-hacking/

product. Finally, we should test whether we have the right price to attract customers to activate, while also maintaining good profit margins for us.

The best way to retain customers is to make sure they are happy and engaged with our product. This is why the Grow stage is not just about scaling the product. Part of the focus should be on improving the product to ensure that we get better and better at delivering value. For example, we should not focus on giving good deals to new customers only. We should also offer similar rewards to current customers. Rather than treating our customers as disposable commodities, we should treat them as members of a community. We should consistently seek their feedback and improve our product in response to this feedback.

These are some tactics we can use to scale our product. What is important is to always remember to apply lean thinking. Do not apply every single tactic at once. Using the build–measure–learn loop, we can systematically test each tactic and how it is impacting our key growth metrics. We must always remember that we are trying to learn about what works. The discipline is to keep learning with data.

5.8 Learning with data

> *Usually the things that see themselves as immeasurable in business usually reveal themselves to much simpler measurements of observation once we learn to see through the illusion of immeasurability* – Douglass W Hubbard[41]

Decision making in the Lean PLC is based on evidence. As such, whenever we deploy our growth tactics we have to use data to track progress. The effort spent in capturing data is worth it if we want to stay lean and ensure we are not wasting time and resources. We should only collect data that allow us to learn. We don't need to know everything, we just need know enough to focus on our growth hypotheses and track the relevant metrics.

---

[41] Hubbard, D.W. (2014). *How to measure anything: finding the value of intangibles in business.* Hoboken: Wiley.

Another common mistake to avoid is not collecting data due to the perceived of difficulty of capturing them. Whatever you need to learn, you should know that pretty much everything is measurable, albeit with variable degrees of complexity. Don't fall into the trap of thinking that what is difficult to measure cannot be measured. If you need to learn something, figure out a way to measure it! Based on the work of Douglass Hubbard, please consider the following factors when deciding what to measure:

1. What is the decision the measurement is expected to support?
2. What is the definition of what is being measured in terms of observable consequences and how does it matter to the decision being made?
3. How do we compute outcomes based on the value of this variable?
4. How much do we know or what is our current level of uncertainty?
5. How does uncertainty about this variable create risk for the decision makers?
6. Is there a 'threshold' value above which one action is preferred and below which another is preferred?
7. What is the value of additional information?

These seven questions can help us discover the key data we need to capture, as well as how much effort to put into capturing these key data points. This allows us to focus our effort where there is a return in value.

Data categories

There are four main categories of data we recommend for the right side of the Lean PLC. These are *market, customer, finance* and *operations.* By examining and correlating these data groups, you will be positioned to make better decisions around strategic and tactical changes that maximise value. Let's explore each of these data categories and explain what you can expect to learn from each.

1. **Market data**

   Market data will help you observe and learn from your business environment at all times. Within these data you should be looking to see how competitors are evolving, what macroeconomic conditions are changing which could restrict or allow you to exploit your product, how channels and partnerships are performing, as well as emerging social and technological trends. Even as you scale your product, keeping an eye on market data will help us stay ahead of the curve.

2. Customer data

As we have already noted, customers should be at the heart of all decisions made. Data collection efforts should focus on how well your product and business performs against customer needs and expectations. The pirate metrics[42] funnel we have mentioned several times is a good way to track customer data. In general, we should be tracking the following key customer data points:

☐ Customer satisfaction metrics/scores (CSAT) or related measures such as net promoter score (NPS).

☐ Customer services should be actively collecting information on common issues or pain points raised by customers over a set time period. The customers services team should also be actively monitoring and managing escalation categories ranging from *critical* to *minor* issues by type, product release versions and volume.

☐ Product performance against the customer job to be done. This could include time on task, retention/repeat visits and number of identified tasks completion.

☐ Time to value from point of sale to customer engagement.

☐ Customer recommendations and referrals to other potential customers.

☐ Lifetime value of the customer by customer cohort and across the group.

3. Finance

It goes without saying that you will need to monitor and capture financial data on an ongoing basis. You will need to monitor actual financials against forecasts across all operations and revenue streams. This should be done on an ongoing basis to ensure that you are on track and are responding to any negative or positive changes. You will need to track cost (including variable costs), revenues and profit margins. Your goal is to create and scale a sustainably profitable business model for the long term.

---

[42] McClure, D. (2010). 'MoneyBall for startups: invest before product/market fit, double-down after.' Master of 500 Hats. Available at http://500hats.typepad.com/

4. Operations

Operational data will allow you to learn how your internal business engine is performing under the bonnet. You can learn a lot about your ability to scale and where performance opportunities exist within your operational data. Some of these may extend financial data and others could be around performance of the business itself. Typical data points include:

- *Frequency of product releases* – Remember the shorter the cycles, the better the propensity to learn, exploit opportunities, adapt to change, provide customer value and eliminate waste.

- *Product availability* – Your service availability is key to customer satisfaction. Monitoring the uptime, downtime and response time of a service can reveal a lot about your product maturity and ability to scale while still satisfying your customers. As you grow, it's likely that your technology or operational stack will be challenged and you will need to know where the points of failure are or are likely to be.

- *Incident monitoring and reviews* – These measure the frequency and type of major and minor incidents across the business. This can stretch across all departments to include customer issues, service issues, response times and other operational issues.

- *Employee engagement* – This can be a revealing data point to observe. As your customers are central to your business model, so are your employees. Employee engagement which could include satisfaction, sickness, retention, vacancy referrals could reveal relationships to other key data points. As your business scales, your culture will be challenged. Don't overlook the value of people and your teams. Great companies who scale recognise the valuable contribution of their employees.

Although reviewing these data points may seem intimidating, many of the data points we recommend can be populated or retrieved automatically through many of the common systems already in place. Many companies beyond a certain size have the systems in place already capturing key data, although it's often inaccessible to product teams. For example, the following systems which many of you may already have in place can be sources of great knowledge to help you learn:

1. ***Customer relationship management systems (CRM)*** – Systems such as Salesforce, Microsoft Dynamics, SugarCRM, Netsuite CRM, Zendesk, Freshdesk or others will likely have a great deal of information on your customer base, customer issues, sales funnels and marketing inbound leads.

2. *Enterprise resource planning (ERP)* – Systems such as Oracle EBS, SAP or JD Edwards will likely have all your financial, HR and operational data stored giving you an overview of costs, revenue and profits for your business model.

3. *Application lifecycle management (ALM)* – Systems which monitor your delivery flow and service status such as Atlassian JIRA, VersionOne, Rally, LeanKit and others contain a wealth of information around operational state and delivery cadence. Such systems may potentially provide insight into flow which includes defect trends, mean time to restore systems, deployment frequency and lead times.

4. *Corporate performance management (CPM)* – Systems which are integrated with many elements of the operational, planning and controls of an enterprise. Solutions such as Anaplan, Host Analytics and Hubble can provide and help release trapped knowledge within your business helping empowered employees make decisions using key data points. Such systems can integrate across data sources and provide enterprise-wide performance insights through data analysis. Releasing this information on a real-time basis could significantly inform your Lean PLC progress and provide key insights to business model design.

5. *Business intelligence (BI)* – Tools which provide integration, collection, analysis, and presentation of business information. Tools such as Tableu, Microsoft Power BI, Qlik and Domo can be configured to present key data trends and KPIs from multiple data sources. Utilised well these can provide beneficial outcomes which could relate to your products performance relevant to your lifecycle stage.

Although many systems have been referenced above, these are *not* required or necessary by any means to operate successfully with the Lean PLC. In fact, for smaller lightweight and agile teams, we would advocate keeping your processes very simple and avoiding ALM tools until absolutely necessary, where the benefits are understood and explicit. Many smaller companies want to appear like bigger companies so are tempted to adopt their practices. This can introduce waste and limit the nimble and responsive states that some smaller companies should take advantage of.

Teams can succeed with simple operational tools and processes, often observed with development teams using iterative development practices using nothing but post-it notes and office wall space supported by the Scrum Process.[43] The important point is to focus on growth and key outcomes in line with your strategic vision and to use the appropriate data when necessary. If you are large enough or operate in an environment with any of the systems above already in place, we would however encourage you to unlock the data trapped within these systems to provide meaningful knowledge to help inform your Lean PLC phase. This explicitly requires you to challenge organisational cultures which may prohibit cross-departmental and cross-functional working collaborative working.

The range of data categories can also be extended or altered as you see fit in order to meet your business needs and domain. For instance, Pearson have a whole data category on learner and efficacy outcomes for all their educational-based products.[44] This supports their global strategy that all products must demonstrate clear benefits to their learners. Tesco have a strategic category on contribution to local communities which is measured using performance scorecards on a store by store basis.[45]

 Tip

As obvious as this may sound, ensure that seasonality is reflected in your growth hypotheses and data trends. For example, the 8-week Christmas period accounts for one third of annual sales in the fragrance industry.[46] Also when measuring more complex lagging indicators such as efficacy or lifetime value of the customer (LTV), this might take more than a quarter. You will need to decide when a KPI measurement will be best measured in reference to time in order to be meaningful.

---

[43] Sutherland, J. and Schwaber, K. (2016). *The Scrum guide.* Available at http://www.scrumguides.org/docs/scrumguide/v2016/2016-Scrum-Guide-US.pdf

[44] Efficacy and research at Pearson. Available at https://www.pearson.com/corporate/efficacy-and-research.html

[45] Tesco on supporting local communities. Available at https://www.tescoplc.com/tesco-and-society/supporting-local-communities/

[46] Blitenthall, M. (2016). *The sweet smell of success.* Available at http://www.nielsen.com/nz/en/insights/news/2016/the-sweet-smell-of-seasonal-success.html

5.9 Metrics are people too

> *Out of all the things we did, the most important change was in how we thought. In simple terms, we reversed the flow of the company: instead of our work ending with the customer, it started with the customer. That basic principle embodied our desire to avoid letting them down again. We would do everything and anything to ensure that we knew what customers needed, and that we gave it to them before any of our competitors did* – Sir Terry Leahy[47]

One of the points that can get lost when exploring data is that there are customers behind each of those metrics. As businesses scale, the focus on improving the numbers can lead to counterproductive behaviour. This is a particular challenge when it comes to handling financial metrics. There is often a lot of pressure to keep those profit numbers improving. Over time this pressure can lead to poor decision making and subsequently long-term loss.

A lot of companies fail to realise that you can improve profits in the short term, while destroying value in the long term. This is because profits are a ratio. They can be improved by revenue increases or by cutting costs. In a competitive market where the cost of switching is low, companies must avoid falling into the trap of increasing revenues by raising prices while not improving on the value they provide. Cutting costs can also lead to increased profit but a poorer service for customers. In a competitive market or a well-defined category, not balancing this ratio with customer value in mind can be costly.

Metrics are people too. Regardless of the Lean PLC stage we are in, the focus on customers must be ever present. As we deploy our growth tactics, we will be tracking our key metrics. However, we should also consistently get out of the building and talk to customers face to face. This is especially powerful when our growth tactics are not working. Instead of just trying different tactics to see what sticks, we should find our key customers and talk to them. The lessons learned will add more colour to our numbers and ensure that we keep delivering value.

---

[47] Leahy, T. (2013). *Management in 10 words*. London: Random House.

5.10 Managing prioritisation through risk and value

No matter where your product is in its lifecycle, the environment, competition and future changes ahead will present risks and uncertainty. To help you adapt, apply and recognise new learnings and challenges that have emerged, we recommend regularly facilitating a prioritisation and roadmap review. These reviews should reflect your experimental outcomes and look to apply the learnings directly. There are a number of different practices and techniques that you can apply to facilitate prioritisation sessions and we have shared just a few below. The important points to consider when reviewing your priorities are:

1. Evidence and data will reduce prioritisation subjectivity. Contribute learnings from previous experiments and surface examples containing antilogs (things that have failed in the past) and analogues (similar things that had worked in the past). Remember internal validation has limited to no value. There is no substitute for direct customer feedback.

2. Re-state the problem you are solving for who and why. Overtime assumptions creep in and can blur the focus or goals previously set. It is beneficial to recap on the goals and outcomes to help re-prioritise goals as well as helping new ideas to be prioritised against.

3. Ensure you have a cross-section of representatives. If you're organising a group facilitated session, ensure you have a small, but decisive group which contains a spread of stakeholder and/or functional roles.

4. Ensure you understand and are clear on the strategic objectives. When we engage in our day-to-day activities, many opportunities surface for pivoting. Although all learnings are captured, it is important to ensure the correlation with strategy is clear as this may affect stakeholder value and continued sponsorship.

With these suggestions in mind, below are some techniques you can apply to help prioritise your backlog.

Prioritising risky assumptions

Building upon the earlier exercises which explore the identification of risky assumptions, we should now have more information to work with being in Grow. Taking a similar approach, the following exercise which is a light version of a traditional risk monitoring approach, can be quite revealing and powerful to facilitate with your team. This is particularly useful when referencing risk as a confidence indicator when planning your product roadmap. To explore new risks and assumptions, the following six steps exercise can help your team prioritise what to do next.

1. **Business model review** – Review your business model canvas and call out and untested introductions, opportunities, risks, changes as well as new learnings. For instance, to grow to the next level you may introduce a new channel or key relationship and may want to test the value of this before committing. You can also review any new inputs and learnings at this point around your business model as a whole.

2. **Learning reflection** – Reflect on your growth hypotheses and learnings to date. Then review your proposed growth hypotheses for the next period 1–3 months.

3. **Capture outstanding assumptions** – With your teams, individually list assumptions, goals and risks.

4. **Group visibility** – Group all the results and list them on a board for everyone to see and input into.

5. **Scoring** – Score each item by the risk of failing and the impact to the business if not true. It is recommended that you establish and define a few examples for each scoring reference. For example, an item which has evidence of high risk of failure or conversely has no evidence to reference should increase the risk of failure to 5. The impact rating should be informed by the dependencies of its success against your business model projections. For instance, if you will cease to operate if regulatory needs aren't met or you will put at risk continued sponsorship, then the outcomes of this should be scored 10.

6. **Refine** – Discuss and refine through conversation as a team.

Table 5.2 Impact of uncertainty scoring example

	Assumption	Risk of failing (1-5, 5 High)	Impact if not true (1–10, 10 High)	Uncertainty/ risk score
A	We believe our new reward benefit will increase referrals by 5%	3	8	24
B	New focused e-learning content on the site will increase product activation by 10%	3	6	18
C	We can integrate 3rd party single sign on within 2 months	2	5	10

By applying evidence from learnings, antilogs and analogues, you are able to highlight a reasonable prioritised list to tackle next with an improved level of understanding. This fairly quick way to numerically quantify some of the risks, helps you prioritise your backlog for experimentation. It will also help invoke interesting conversations as a team, surfacing further insights whilst providing improved team alignment. You may be surprised to see what new evidence surfaces when facilitating these sessions.

Cost of delay and weighted shortest job first

Depending where you are on your growth path, another factor to consider is the cost of delay. This is useful where you have a lot of ideas channelled through a bottleneck competing for resources and need a logical prioritised path of value. When your business model has a level of increased certainty you will be able to call upon more evidence and lean more towards operational efficiency. This is where the weighted shortest job first (WSJF) can be beneficial as a prioritisation effort, especially considering the effort versus returned value.

> ## Q Tip
>
> Don't make the mistake that many do by assuming you're in a predictable state where assumptions are confused with facts. We recommend regularly reviewing your riskiest assumptions for the whole period that you're in Grow. Therefore, however you prioritise, always consider data and evidence as a prioritisation qualifier. Where there is an absence of data and evidence, consider prioritising and obtaining this first. This should be a small investment in time and will reduce risk significantly.

The WSJF is a common practice advocated by larger agile development teams to capture value and prioritise the product backlog. This is further explained by Donald G. Reinertsen in his book *The Principles of Product Development Flow.*[48] This approach considers the following three aspects when prioritising the backlog.

1. **Shortest job first (SJF)** – The principle is to do the shortest job first where the jobs have the same cost of delay. If one were to do the longest job first, it would delay many short jobs and result in the 'Convoy Effect', delaying the overall time to value.

2. **High delay cost first (HDCF)** – Where job durations are homogenous, but the cost of delay differs, it's best to do the high delay costs first.

3. **Weighted shortest job first (WSJF)** – When both duration and delay costs are different which is most common, the best approach is WSJF. Priority is based on delay costs divided by task duration. An example of this can be seen in Figure 5.2.

As further highlighted by Reinertsen, there are three common mistakes to avoid when implementing this process.

1. Assuming the return on investment (ROI) alone should be used for priority where high ROI projects are prioritised over lower ROI projects. You should consider the sequence of the ROI and the sensitivity to profit. A project with lower ROI, but higher

---

[48] Reinertsen, D.G (2009). *The principles of product development flow: second generation lean product development.* 1st edition. Celeritas Publishing.

Figure 5.2

Feature/Project	Duration/Effort	Cost of Delay	Weighting =(COD/Duration)
A	4	8	2
B	5	7	1.4
C	5	6	1.2

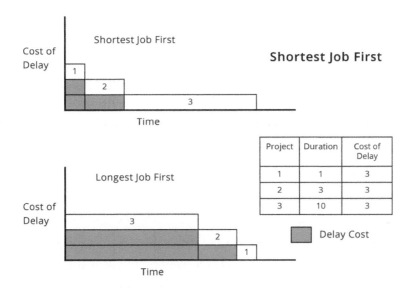

sensitivity to profit should go first. Overall portfolio ROI adjusted for delay is more important than an individual project ROI.

2. In the absence of quantitative information, companies can operate on a 'first in first out' basis. Although simple, this is dangerous when jobs have a different delay cost.

3. Some companies use a critical chain which allocates buffer time to each job and gives priority to the project with the smallest remaining buffer also known as minimal slack time first (MSTF). This aims to minimise the total number of projects that miss their planned schedules. This approach is not optimal when considering overall delay costs and projects with a higher cost of delay with ample slack time should take priority over projects with lower costs of delay, but with less slack time.

When taking a hypothesis-driven approach to calculating or estimating the WSJF, we would strongly advise that the values are informed by evidence. When working with some teams and particularly those in larger enterprises with reference to mature products, we advocate using evidence as a multiplier. This is particularly important if you're operating within a *complex* or *complicated* domain as defined by the Cynefin framework'[49] where discovery is expected. Evidence helps calculations of WSFJ to have a higher degree of certainty, removing some subjectivity from the scoring. Taking the example shown in Table 5.3, you can see how this can transform the previous priorities set.

Table 5.3   WSJF – evidence as a multiplier

Feature/ Project	Duration/Effort	Cost of delay	Weighting	Experiment evidence level	Confidence value
A	4	8	2	0.2	0.4
B	5	7	1.4	0.8	1.12
C	5	6	1.2	0.6	0.72

To support the inclusion of evidence in a more structured format, we would recommend that you establish an agreed understanding of weighting towards the evidence provided using a confidence indicator. The more evidence and data provided the more confident you should be with your projected outcomes. Recognising this can help facilitate the emergence and realisation of risk which could reduce waste early. If something is deemed as high value for growth, then supporting evidence should be strong.

You should extend and contextualise your scoring system for your business and environment. It's important that this is explicit and shared before any scores are applied. An example of such a weighting can be seen in Table 5.4.

The scoring guide is an example of how evidence can be considered. We weighted the cost of delay against effort equally to balance the investment to the outcome. The value of the confidence score is that it challenges the commonly observed practice

---

[49] Wikipedia (2017). *Cynefin framework.* Available at https://en.wikipedia.org/wiki/Cynefin_framework

> ## Q Tip
>
> Hierarchy is built into the cultural operating system of society and does influence behaviour and direction whether we like it or not. The consequence of this is that ideas born from HiPPOs (Higher Paid Person's Opinions) are seldom challenged or refined to be more meaningful. This isn't necessarily the fault of the individuals but is inherited through our cultural development of behaviours, experiences and the working system. Where failure isn't accepted, few people like to take high risks, especially in situations where accountability is at play. In almost all our experiences dealing with HiPPOs, we have come to learn that the favourite food of the HiPPO is data! We actively encourage the collection and contribution of data for any proposed ideas and recommend you feed the HiPPOs!

Table 5.4 Confidence value scoring

Evidence type	Comments	Score
Legal & regulatory conditions	Where you need to apply regulatory or legal conditions to operate which are non-negotiable, then the maximum score is assigned by default.	1
Research sources	Cost of delay	Max value 0.5
Market research	What worked for others doesn't mean it will work for you and your customers. Market research is valuable, but not absolute. Market research should provide insights and depth in addition to experiments. Consider that just because market experiments are successful, it doesn't mean the market is big enough to warrant the investment.	<=0.2
Market experiments	Market experiments have been carried out with real customers and consist of enough of a sample set. The data demonstrate clear customer intent to commit to the solution on a repeatable basis.	<=0.3

Table 5.4 Continued

Evidence type	Comments	Score
Development sources	Duration/effort	Max value 0.5
Previous development experience (3 parts below)	Estimates are best guesses at a given point in time, often with limited information provided. Therefore, it's recommended this is recognised to help indicate confidence. Low confidence can trigger exploration and experimentation which could rapidly improve confidence. If confidence is low for effort estimation, then consider running short, focused experiments first.	
Dev experience (High)	Where you are using the same team who have developed very similar or the same components in the past many times and they have directly contributed the effort estimate with supporting evidence and references. The team may also reuse many existing components and patterns from previous implementations, prototypes and experiments. Best practice can be applied.	<=0.5
Dev experience (Medium)	Where you have team with a high ratio of new members or members who haven't worked together much previously and only can reference 1–2 similar projects with limited direct evidence. They only have some pre-existing resources to utilise and will need to explore the difference. This could be considered complicated, but good practice can be applied.	<=0.25
Dev experience (Low)	Where you have a new team and the challenge is new to all or most members, where little or no resources, knowledge or previous patterns can be utilised. This could be considered a complex state and therefore presents higher risk. Discovery and experimentation is encouraged.	<=0.1

where people treat estimates and numbers as fact. That's not to say there is no value in estimation, but the added realisation of risk and confidence will encourage teams to experiment more.

Combining risk and value

Building on the technique above as an alternative to the confidence indicator, another useful exercise is to combine both the values above: impact uncertainty scores and WSJF. Whereas the WSJF in practice can sometimes omit uncertainty, the uncertainty score can omit the effort and duration and the benefits of WSJF. Where you have higher risk features, ideas, projects or stories it would be beneficial to experiment and obtain more knowledge before deciding to invest significant effort and detail. Taking the Risk prioritisation and the WSJF technique above, it can sometimes be beneficial to visually cross-reference the values. The following model illustrates this point.

Value-uncertainty prioritisation quadrant

In Figure 5.3 we are recognising that although there equally or similar weighted scores for both Risk, Impact and WSJF, when combined there's a distinction which realises limited certainty and/or understanding. The top right quadrant recognises that there

Figure 5.3

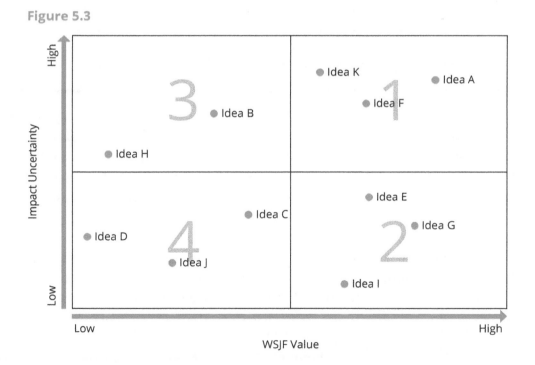

are three ideas which present a higher risk of uncertainty against higher value of return, measured by WSJF. One could argue that these ideas would need to be prioritised and explored further to reduce the risk which might position them in the lower right quadrant. Recognising this explicitly surfaces questions and uncertainty back to the stakeholders to not yet bet on these outcomes. It also implies more exploration and knowledge is needed before value can be realised.

If prioritising the ideas above, the order of priority recommended would be to follow the number convention from 1–4 as a guide. Some subjectivity is required as the metrics informing the positions in the first instance are subjective. Taking the example of idea H against idea C, you might prioritise idea C first as the cost of delay in the WSJF may have more value than the risk exploration than idea H.

 Tip

It's important to note that evidence and inputs should be used to support the numerical projections above. Estimates alone without qualitative evidence to help quantify your backlog of ideas will limit the value of prioritisation.

5.11 Managing a nonlinear adaptive state

With all the best projections and plans, in each Lean PLC phase you are highly likely to be in multiple states simultaneously while you expand your business and execute. In the digital economy, nothing stays still for very long and therefore it's important to recognise the constant need to adapt and respond. Not only are you looking to make unknowns become knowns, but your environment could also reduce the lifespan or relevance of some knowns over time.

Providing insights into understanding your domain state and how to operate successfully to build successful products in this space can be established through the support and reference of the Cynefin Framework.[50] This framework identifies five domain states.

---

[50] Snowden, D.J. and Boone, M.E. (2007). *A leader's framework for decision making*. Available at https://hbr. org/2007/11/a-leaders-framework-for-decision-making

1. **Simple/obvious** – Represents known-knowns where the relationship between cause and effect are clear. This domain advocates best practice through 'sense–categorise–respond' behaviour.

2. **Complicated** – Represent known-unknowns where analysis and/or expertise is required to understand the cause and effect relationship and then apply appropriate good practice. The framework recommends 'sense–analyse–respond' behaviour.

3. **Complex** – Represents unknown-unknowns. The cause and effect relationship can only be understood in retrospect within this domain. The framework recommends 'probe–sense–respond' behaviour. This can be achieved through safe to fail experimentation.

4. **Chaotic** – The cause and effect relationship are unclear and as such a knowledge-based response cannot be applied, and any action is the most appropriate initial response to establish order. The framework recommends 'act–sense–response' behaviour which allows the responder to use the established order to then sense and better understand the situation then respond accordingly which will move the domain to being *complex*.

5. **Disorder** – Unclear which domain applies and there are multiple perspectives in conflict. It's recommended that event or situation is dissected into parts and distributed appropriately to each of the other four domains.

This sense-making framework provides much depth and realisation into the preferred paths and ways of working when you identify your product and business model context. It is also possible to be in more than one state at a given time depending on what level of granularity you observe. For example, your customer services team could be operating within the *simple* domain, while R&D operate in a *complex* domain and your business model across all domains across its respective areas. It's worth noting that your domain state can change over time with environmental influences. This is why we recommend reviewing your strategy, vision and tactics on a regular basis.

The common overall domain state of product development when in the Grow stage of the Lean PLC is often likely to be weighted towards the *complex* and *complicated* domains. This is due to the inherent complexity of product development and management, coupled with emergent complexity and knowledge around your business model and environment in Grow. These domains reveal more than one right answer to any given problem or hypotheses and advocate the recognition of assumptions to be explored and validated. Recognising which state you're in will help inform preferred operational practice. The realisation that product development in the Lean PLC phases from Idea through Grow is *complex* to *complicated* is recognition that you're exploring known-unknowns and unknown-unknowns; where the learning in the latter state by definition can only be obtained in retrospect.

A common mistake that businesses make is assuming they are in a *complicated* or *simple* domain where there are fewer unknowns. They operate where exploration, discovery and change are unwelcomed or not expected and then default investment to longer term plans which are highly detailed. The Lean PLC on the contrary advocates emergence and continuous learning, particularly through Idea to Grow. This supports the behaviours of probe–sense–respond (*complex* domain) or probe–analyse–respond (*complicated* domains) to improve the chances of success and maximise learning. This is why the similarities of build–measure–learn practices and the scientific approach of hypothesis-driven development can be exploited to maximise your potential from Idea through Grow Lean PLC phases.

5.12 Capturing your product roadmap

Your product roadmap is an extremely powerful and a living artefact which impacts every function in your organisation. It's a declaration of all the knowns and unknown outcomes ahead, built on the presumption of the most valuable return on the investment of effort contributed. Don't get misled by the term 'Product'. The Product Roadmap is the path of value delivery which an entire business is undertaking. The ongoing delivery of the product is the repetitive transaction of value between the customer and business. Your roadmap is the one true reflection of all the points previously discussed, disseminated into a clear and transparent sequence for all to see and contribute towards.

In order for the Product Roadmap to be most effective, there are some common attributes which should be sought and employed. An effective roadmap should be simple and need little or no explanation to those consuming it. The following list of properties should be considered when forming and supporting your product roadmap.

1. **Prioritised:**
   (a) What have we completed?
   (b) What are we currently working on?
   (c) What is coming next and why with clear strategic alignment?

2. **Engaging** – Clearly articulating value and supporting activity undergone and to be discovered ahead against a clear goal and purpose of what and why something is being delivered or achieved.

3. **Accessible** – Roadmaps should align product organisations and promote engagement. Therefore, they should be accessible and available to a wide audience and in a language which requires little or no explanation.

4. **Up to date** – There should be one single source of truth which is up to date. A more up to date roadmap will likely encourage more engagement and contribution.

5. **Strategic** – The product roadmap is a strategic artefact which communicates how you are delivering on your strategic objectives. Stakeholders should be able to clearly understand how the product roadmap seeks to deliver value against the strategic goals set.

Although a single roadmap is desirable, in larger organisations it's likely that you will need to expand the details and provide different views of granularity for the different viewers. As companies need to be more adaptive to change and are becoming learning organisations, using tools which are more open to discover value and explore outcomes and respond to change are more effective for the Grow phase of the Lean PLC. A useful tool to use is the Go Roadmap by Roman Pichler.

Go Roadmap

The benefit of this approach is that the Go Roadmap is goal orientated which allows teams to inspect and adapt as they delivery to shape to best solution for the desired outcome. Supporting the explanations during Explore, Validate and Grow, the Go

Roadmap also supports a hypothesis-driven approach with key measurements and metrics for review. It's also at a high enough level that it can be shared across a broader range of your organisation.

Figure 5.4[51]

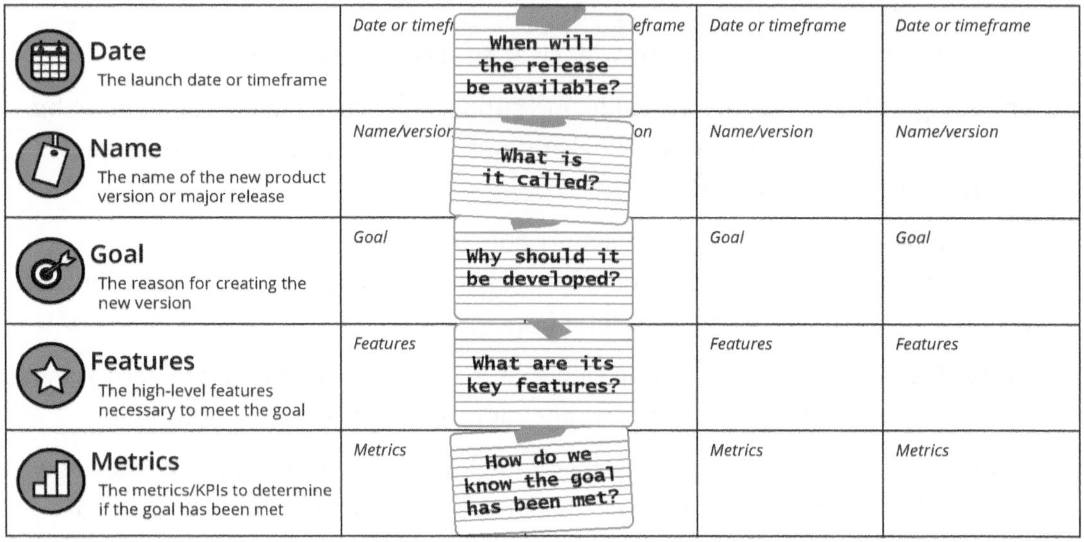

	Date or timeframe		Date or timeframe	Date or timeframe
📅 **Date** The launch date or timeframe		When will the release be available?		
🏷️ **Name** The name of the new product version or major release	Name/version	What is it called?	Name/version	Name/version
🎯 **Goal** The reason for creating the new version	Goal	Why should it be developed?	Goal	Goal
⭐ **Features** The high-level features necessary to meet the goal	Features	What are its key features?	Features	Features
📊 **Metrics** The metrics/KPIs to determine if the goal has been met	Metrics	How do we know the goal has been met?	Metrics	Metrics

If you are working in larger organisations need to communicate intentions and state across many locations and people, you may want to use a digital tool. There are many options available and the choice of tool should be carefully selected not to compromise the principles expressed in the Lean PLC. Using a tool shouldn't replace the behaviours and working practices advocated, but should be used to complement them and to add value.

A useful roadmap tool which we have used with some teams is http://www.aha.io, which offers a range of solutions which support goal orientated product roadmaps with references to business model. Another powerful capability of this tool is that it can translate and provide different views and granularity levels of the same path. Having multiple views presented from one single source of truth can be useful.

---

[51] Pichler, R. (2016). *Strategize: product strategy and product roadmap practices for the digital age.* London: Pichler Consulting.

Whatever roadmap tool or technique you use remember you employ a tool to work for you and not the other way around! The practices shared during this book are advised as paramount over the use of any tool chosen. A Gantt chart of any kind is not a defined path of the future, but a shared snapshot of understanding at a given time. Another way to look at this is that the roadmap is simply a list of questions yet to be clarified to achieve the value desired.

A useful tip employed by many teams is to explicitly communicate certainty on the roadmap itself to ensure that the roadmap doesn't end up being a contract of delivery. An example of this is shown in Table 5.5.

Table 5.5   Product roadmap implementation maturity

	1–3 Months	**3–6 Months**	**6–12 Months**	**12+ Months**
Risk of change	Low	Subject to Change	Very likely to change	To be defined
Features	These features should have clearly defined measures of success aligned to objectives and goals which are regularly tracked for progress and delivery status.	Features at this stage are being refined. Some exploration and detail added over time. The closer the feature gets to implementation, the more understanding surfaced.	High level definition of what value is sought and problems to be solved, captured at an epic/idea level.	Ideas can be captured and grouped for review against strategic goals. Granularity is very high level at a research and direction level.
Granularity	Features, stories, tasks, experiments	Epics, features, experiments	Epics, ideas, goals, experiments & research	Strategic goals, objectives & research

Table 5.5   Continued

	1–3 Months	3–6 Months	6–12 Months	12+ Months
Questions	Continuously updating delivery throughput and cadence. New learnings through application and delivery of features.	Questions and experiments that are being explored and explanation as to what value they aim to provide with the risks to mitigate.	Epic level stories should be looking for evidential information which will likely require exploration and further research.	Questions should be less about features and product capabilities and more about the directional capabilities needed.
Comments	Features have strong supporting evidence and definition informed by previous exploration. It is believed there are more knowns than unknowns at this point with the exception of experiments being delivered.	A strong connection should be made between the current activity and this phase. The learning outcomes from the stages 1-3 months should be continuously applied here and vice versa.	Epics at this stage are expressions of ideas. Epics should be prioritised and managed using techniques previously expressed in the Lean PLC. Path-finding efforts should be looking to change unknowns into knowns.	In previous chapters we encourage you to explore outcomes from tactics through quarterly reviews. These should inform your vision and strategy.
Update cadence	Daily/Weekly	Bi-weekly/ Monthly	Monthly/ Quarterly	Quarterly

5.13 Entering new markets

Another way to accelerate and expected growth is to enter new markets. Common questions surface when mapping products to a stage in the product lifecycle around whether a product should be mapped by distinct markets. As a general rule of thumb, new markets present new and sometimes very high risks and differences which should to recognised. There are countless cases where very successful products and businesses in one market dramatically fail are constrained or choose not to operate in distinct markets.

By way of example, the tech giant Google who have well publicised success on most global regions, pulled out of China in 2013, despite the Internet user base of China being twice the residency population of the USA.[52] Another retail and tech giant Amazon which is no doubt the default online destination for purchasing goods and maybe the distribution channel you purchased this very book from, has only recently entered and expanded in Brazil.[53] This is due to the unique challenges and difficulties faced in this market place. Both these giants carefully made their decisions knowing the growth model in each of these markets is unlike the other markets they have success in and needed to refine their business model or review the strategic impact.

Wherever you are in the lifecycle, when entering a new market we would advocate that you start from Idea to explicitly recognise there are unknown-unknowns which will need discovery and validation. This doesn't mean that the product shifts its stage in the lifecycle, but simply means that each market can be in a different stage. You can have a product in more than one product lifecycle stage at a given time. Let's say for instance you are in Grow and you have a product in the USA, but you want to go to Brazil. Your product overall and a high level would be in Grow (total combined growth), but in one market it could be in Explore (Brazil) and another in Grow (USA). Let's take a look at the following example:

In Table 5.6, the product and business model is being refined and contextualised per market, but overall is considered to be in Grow. Collectively this declares that the product and business is expected to see growth. Market contextualisation is a key

---

[52] The Atlantic –Waddell, K. (2016). *Why Google quit China—and why it's heading back.* Available at https://www.theatlantic.com/technology/archive/2016/01/why-google-quit-china-and-why-its-heading-back/424482/

[53] Bloomberg-Moura, F. and Sciaudone, C. (2017). *Amazon expands in Brazil, making worst-kept secret official.* Available at https://www.bloomberg.com/news/articles/2017-10-18/amazon-expands-in-brazil-making-worst-kept-secret-official

Table 5.6 Mapping markets to Lean PLC

Idea	Explore	Validate	Grow	Sustain	Retire
			Overall product		
Brazil	India	Australia	UK	USA	China
		Spain	Japan	Canada	
			Germany		

factor in success and which speaks to the reasons why Amazon may have taken longer to expand into Brazil.

What defines a new market?

1. A new geographical boundary where your product does not yet have a presence. If you have a product which is serving a market, such as the USA and you would like to start selling in the UK, this would be defined as a new market. The UK entry should start from Idea and evolve through the Lean PLC stages utilising the governance and principles provided.

2. A new market, could also be a new demographic in an existing geographic boundary. If you currently serve a particular user base such as consumers in the USA and now you would like to serve business users in the USA, this would be considered a new market.

Q Tip

If managing a product portfolio or multinational product in the Lean PLC, you can change the lens by market and the collective view. In fact this is strongly encouraged in order for you to not make the explicit separation by market, and reducing the risk of wrongly assuming that the current business model need not be refined. In each market your product could be in a different Lean PLC stage, but overall if you are growing your market share it could be collectively in Grow. From a financial perspective in the portfolio, you can set growth hypotheses per market and overall for the product as a collective. Generally this reduces the risk of financial investment by assumption.

A case study: Monzo in Grow

An example of company executing in the beta phase of Grow at the time of writing is the disruptive FinTech company Monzo (previously known as Mondo). This rapidly growing company which raised its first £1m in just 96 seconds via crowdcube.com in March 2016,[54] is growing a bank in beta and could be referenced as being at the early stage of Grow in the product lifecycle.

Distributing approximately 1000 debit cards per week and managing a customer waiting list of 52,000 people, Monzo clearly have found a value proposition which is attracting huge interest. They also have to grow at a sustainable rate they can support, while keeping customers interested and engaged in a very competitive environment.

Monzo have extended their capability to scale by harvesting a community knowledge base at community.monzo.com. Monzo's community strategy is a very tactical growth programme, helping Monzo to accelerate and learn fast. Members suggest product improvements, report issues and even support each other in solving issues that arise creating a continuous cycle of improvement with validated customer input. This gives Monzo a lot of insight into what is the next most important thing to develop and improve and where there is huge value to be obtained. In August 2016 they obtained their much-anticipated banking license. As such they are required to mature their business processes and technologies to improve compliance and risk management to satisfy regulatory authorities and third-party interfaces. Such practices will likely be expected from the early majority and beyond where higher quality and standards are expected for product adoption. Their continued customer expansion is matched by their ability to continuously learn.

---

[54] Mondo reaches £1m in 96 seconds in record breaking raise | The Crowdcube Blog. 2016. *Mondo reaches £1m in 96 seconds in record breaking raise | The Crowdcube Blog.* Available at http://blog.crowdcube.com/2016/03/03/mondo-reaches-1m-in-96-seconds/

5.14 Prepare for sustain – when growth slows down

Our hope is that our product grows and continues to grow for many years. But nothing lasts forever. At a certain point, the product will reach a level of maturity in the market and growth will slow down. Recognising whether you're in Grow or Sustain is important for Lean PLC management. This helps align expectations and goals so that appropriate resourcing, investment, support and revenue projections can be considered.

It is also important to note that, unlike the Lean PLC stages on the left side, moving a product from one stage to another on the right side is not based on teams making the decision and bringing the product to the council. When a product is on the right side of the PLC, it may remain in Grow or Sustain for many years. As such, Lean PLC governance on the right side is done via quarterly, bi-annual and annual reviews. The cadence of the reviews depends on the size of the company and the business models of each product.

At each review, the product council or investment board can analyse the performance of the product and make decision. Below are some key points to consider during these reviews in terms of making the decision to move a product from Grow to Sustain.

1. **You have had 3–6 review cycles in Grow which have revealed limited or lower-than-expected outcomes in total revenue or customer numbers.**

   It is useful to set specific margins as thresholds to act as triggers for stage specific strategic conversations to emerge. When reviewing the performance and thresholds from Grow, some points to consider:

   (a) If you have a large market share, smaller margins of total market growth can be significant. For instance, 5 per cent growth for a multinational or market leader can result in millions or billions in revenue. Also setting thresholds contextually to your industry and company is key for stage positioning.

   (b) If you are smaller company or a larger company in earlier stages of Grow, it would be advised to apply larger margins as your thresholds, but this is in reference to your domain and goals.

(c) In reference to the margins above, you'll need to look at the average over a period and not react without considering seasonal trends.

If considering all these factors you see consecutive decline, then strategic conversations should emerge to understand why and how to react. Sometimes this results in jumping from Grow to Retire, bypassing Sustain altogether. It's important to note that lack of growth, could be as a consequence of tactical effectiveness and not the lack opportunity. Therefore, understanding all of the learnings to date should form the basis of stage movements.

2. **Your corporate strategy has shifted and you're investing in other products in Grow.**

   This change can mandate your product's Lifecycle stage, irrespective of your trend data. Changing strategic priorities and investments might mandate a move from Grow or Retire into Sustain. Remember the PLC is not linear.

3. **Environmental and political factors limit growth.**

   Consider pharmaceutical or energy companies who are sometimes regulated to maintain stock level and production flow for a set period of time. This can impact the lifecycle stages of their products.

The move from Grow to Sustain raises one fundamental question: How can we maintain revenues and profits while reducing costs? This also raises fundamental product roadmap questions. As such, as we prepare to move to Sustain we should be able to answer the following questions.

1. How can you reduce your team size and other operational costs?

2. Does your roadmap focus on maintaining customer numbers, satisfaction and engagement? Is your roadmap weighted too heavy on discovery over execution?

3. Entering new markets would automatically position your product into Grow. By declaring your product in Sustain, you may be declaring that you are not attacking new markets as defined in Grow.

4. If you're a digital company, how will your product keep pace with the third-party delivery platforms? Consider the frequent releases operating systems, user interface

(UI) capabilities and evolving application programming interfaces (APIs) that your application might rely on.

5. How will you defend market share and what indicators will you need to monitor for early signals to respond to? How quickly can you react to change?

With these captured and defined, you'll need to monitor and manage the transition from Grow to Sustain. This rarely happens quickly and should be handled and coordinated to assure you are seeing the impact of changes and that you're limiting the negative effects on the company. Remember that products in Sustain should be profitable with optimised margins and operations.

SUBMISSION TEMPLATE
Grow to Sustain SUBMISSION

Product ownership

Investment board	
Business sponsor	
Product owner	

Idea overview

Product name	
Idea description	
Strategic fit	

Product performance

Please describe the financial performance of the product to date.

	Revenue	Costs	Profit
Forecast			
Actual (or re-forecast based on actuals			

Growth lessons learned

Provide an overview of the growth hypotheses that you have tested since the last review and the lessons you have learned.

We believe that:	
To verify that we:	

And we measure:	
We learned that:[55]	

Unexpected learnings

Provide a summary of the unexpected learnings you got during Grow.

Sustain hypotheses

Provide an overview of the assumptions about Sustain that you plan to test in the next stage.

We believe that:	
To verify that we:	
And we will measure:	
We will know we are right if:	

Financial projections

Since you are asking to move to Sustain, please complete the financial data below.

Expected revenue over 3 years	Expected costs over 3 years	Profit margins over 3 years (%)

Resources and funding requested

To Sustain the product, we are asking for (e.g. **dollars, time, people**):

[55] Osterwalder, A., Pigneur, Y., Bernarda, G. and Smith, A. (2015). *Value proposition design*. New York: John Wiley.

* Please note that you also have the option to remain in Grow, go back to Validate or move straight to Retire. If either of these options are what you are choosing, then you need to adapt this template so that you can update the product council on work done so far, key lessons learned, what you plan to do next and the resources you need.

An interview with:
Marc Abraham, Chief Product Officer at Settled

Marc Abraham is an experienced product management practitioner, who has worked for a large number of successful digital organisations – at 7digital, carwow, Beamly, World First, notonthehighstreet.com and currently at Settled. Marc writes about his learning and digital trends on his blog at marcabraham.com. We caught up with Marc to find out what lessons he has learned in his experience of working to grow successful companies from idea to scale.

Please tell us a bit about your experience with scaling products. How tricky can that be?
It's funny, because I sometimes hear people say that reaching product–market fit is the hard part, and that scaling beyond that point is the easy bit. I have experienced first hand how tricky the scaling of a product can be, and it doesn't matter whether you are working at an early-stage startup or a big corporate. I guess the mistake people sometimes make is that it isn't just about scaling your product, it's about scaling your *business.*

That is interesting. Please tell us about some challenges teams can face when attempting to scale and how they might overcome them.
From scaling numerous products, I have learned that 'market-fit', being in a good market and satisfying that market, is far from enough. Suddenly competition starts to erupt from everywhere and the things that brought you success initially are no longer sufficient. This is why:

Challenge 1 – Scaling the back end of your product: I once worked at a company where the engineers had their own internal nickname for the technical infrastructure which underpinned their product. Trust me, it was not a flattering nickname! In their effort to get to market-fit as quickly as possible, they had made lots of shortcuts in building the technical platform with elements that were not reusable. As a result, building on the successful, first version of the product was very time consuming; even the smallest of product changes would take require a month on average!

My suggestion here is to have a wider plan in which your MVP fits into. I strongly recommend not treating your MVP as a one off, and 'just throwing something out there' to see if it resonates. Instead, treat your MVP as part of long and successful sequence of product iterations. I often compare it with wanting to have a number of special marbles in a jar. If that is your desire, I suggest you start working backwards to figure out how you get your first marble and how to go from there. The same applies to delivering value to the customer; you can wait till you can deliver lots of customer value in one go, or you can unlock value early and often.

This approach forces you to consider the reusability of your first product iteration - both from a front and a back-end perspective. This approach will enable you to make well informed scaling trade-off decisions right from the outset.

Challenge 2 – Compliance is part of a grownup product: I hold my hand up for releasing a product iteration which breached some fairly significant data laws . . . Luckily, we found a solution very quickly and avoided causing any damage to our customers. However, it taught me a valuable lesson that whilst it might be 'fine' to not worry too much about complying with data and security legislation when you have got 10 users, that becomes a whole different story when you have got 10,000 users!

What I've learned from this error, is that by including 'compliance' in your definition of 'DONE' of a product or feature, you're less likely to treat data or legal compliance as an afterthought. For example, I am currently working on enabling homeowners to enter their property details online. Even though the current focus is on launching an MVP, we are already thinking about how the home data will be stored securely and protected.

Challenge 3 – Customers expecting more and better: I once worked on a project where we were creating a 'second screen' app, which offered TV viewers with extra content related to whatever they were watching on TV. When we initially launched the app, our users were happy with its functionality and content. However, over time, our user needs evolved: instead of using the app for additional content during the airing of a show, people started using the app before and after the show, interested much more in chatting with other people about the show that they had just watched. I feel this a great example of a product that had reached product–market fit but then needed to evolve with user needs.

I know it sounds obvious, but I've found 'curiosity' to be the best antidote to thinking that you have figured out what customers want and that you simply need to iterate on the product that helped you achieve product–market fit. User needs and problems aren't static and you need to be constantly learning about the customer. Don't get me wrong, I'm not suggesting you should go into 'analysis paralysis' mode. Just speaking to 3–5 (prospective) customers per week, will give you a wealth of information and new insights which you can feed into the ongoing development of your product.

Challenge 4 – Revenue growth needs to continue: When you're trying to get to product–market fit, it's easy to fall in the trap of focusing more on market share than on a sustainable revenue model or profit margins. For example, I worked at a company once which had grown rapidly on the back of a product with very small profit margins. As the market started becoming much more saturated, we needed to move quickly and evolve both pricing model to make sure that we could continue growing the business in a sustainable manner.

Not forgetting about your business model, sounds obvious I know. But I've learned how the hard way that you can get very closely to precipice with your business if you haven't at least thought through a high-level strategy of how you're looking to make money beyond the MVP and product–market fit. For example, whenever I have good idea for a product or a problem I want to solve, I force myself to take a step back and ask myself 'Are we going to make money off this? If so, how?' Even if you don't have all the answers at this stage, that's fine. It will at least give you a direction of travel with respect to your overall business model and value proposition.

Thanks for sharing your insights. It has been very helpful. If you had to summarise some of your learnings regarding scaling, what would be your one final thought?
To quote the great Kathy Sierra, 'Don't make a better [X], make a better [user of X].' Because of some of the scaling challenges that we talked about, there's a risk of neglecting some of the aspects which brought you market fit in the first place. It's easy to introduce a slightly worse customer experience, just because you want to generate more revenue, for example. Even if you might reap some benefit from this approach in the short term, I don't believe it will help scale your product and business in the long term.

In order to not fall prey to this trap, I strongly recommend having a high-level product strategy from the outset. Naturally, this strategy is likely to evolve and that's absolutely fine. If anything, thinking strategically right at the beginning will urge you think about scaling and will give you an initial path to bear in mind, well before you've achieved product–market fit.

6

Sustain

6.1 Welcome to Sustain

Well done! Nice work! You are here because your product has reached a good level of maturity. In Sustain your product is now a cash cow. This means that you are consistently generating revenue and you have a repeatable and well-understood business model. Being in Sustain, doesn't mean pulling back because you have reached the highest point of market saturation. In a globally competitive market, sustaining great operational margins and efficiencies, while continuously delighting the customer have to become more your focus.

Even as we enter Sustain, we must remember to continue using Lean thinking and the build–measure–learn loop from the previous stages of the Lean PLC. The discipline and rigour remain the same, it is just the focus of our practice that is different. In fact, although we advocate operational efficiency, it's expected that to improve and maintain efficiency you will continually experiment and learn. The main change is the strategic context of your product and business model.

In most cases, to be in Sustain means you have reached a dominant market saturation point. Your product is satisfying and attracting the late majority and laggards while holding onto your dominant position. This is an attractive space to be in and there are a lot of businesses looking to eat your lunch. You will need to defend your space and keep pace with your customers' evolving needs in order to continue to stay relevant.

Although reports on product management popularise the excitement of products being in stages where they grow and innovate, successful and profitable enterprises are those that know how to manage and exploit Sustain well. Successful products in Sustain often provide the capital and opportunity for continued innovation across the enterprise. Therefore, doing this well should increase your potential for future innovation; providing you have a balanced and well-managed portfolio of products: The Sustain stage can be split into the following activities:

1. *Maintain the engine*: This involves maintaining market share, customer numbers, revenues and profits. In Sustain, this is the key question for every product review: Are maintaining or losing market share, customers, revenues or profits?

2. *Keep customers happy*: This involves ensuring customer satisfaction with our product. Although we are not making major changes during Sustain, we should be consistently making changes and updates to our products to meet customer needs.

3. *Optimise operations*: This involves systematically reducing costs, improving efficiencies and rationalising our operations. We also have to maintain sustainable relationships with our key partners.

4. *Prepare for retire*: Every product has a lifecycle and will eventually have to be taken out of the market. At some point, there will be significant declines in customer numbers, revenues and profits. Changes in technology and customer preferences can render our product obsolete. The goal is to systematically manage the retirement of our products. This is different from having zombie products or being surprised by a sudden loss of profitability.

6.2 Reduce costs while maintaining good customer service

Although we advocate lean thinking at all stages of the PLC, Sustain is where minimising waste is most explicitly the goal for internal operations. With your product in Sustain, there is a great opportunity to maximise profits, without compromising the value delivered to the customer. Focusing too much on the former and not the latter will likely end in greater costs, as you will see a decline in customers over the longer period.

Product management and performance in Sustain requires a systems thinking approach where repeatable processes can be applied, monitored and managed. Management in this space advocates a lean approach where changes applied internally and to the product and the effects are observed through experimentation.[56] This scientific approach isn't limited to a business model of discovery and growth, but to all changes which can be measured and/or observed for any part of the system.

---

[56] The W. Edwards Deming Institute – Hunter, J. (2013). *Deming 101: theory of knowledge and the PDSA improvement and learning cycle.* Available at https://blog.deming.org/2013/12/deming-101-theory-of-knowledge-and-the-pdsa-improvement-and-learning-cycle/

When considering what areas to focus on during Sustain consider reviewing the following areas:

Where to find optimisation opportunities

1. **Identify, measure and optimise your workflows and processes.** Where you can identify process, look to optimise flow and consider the five principles of Kanban:[57]

   (a) visualise the workflow
   (b) limit work in progress (WIP)
   (c) manage flow
   (d) make processes explicit
   (e) improve collaboratively

   In reviewing your product and business model consider cycle times, lead times, queuing and work in progress limits.

2. **Review purchasing procedures, re-evaluate suppliers and look for cheaper partners.** Be mindful not to compromise quality in this process and remember you're optimising for the whole. Choosing a supplier on the other side of the world might reduce your unit price, but could increase your total stock carried due to the logistical lead times. This choice could also affect your corporate strategy as you may negatively impact your company's commitment to reducing your carbon footprint. You should also consider that the quality of interaction and knowledge transfer can also suffer with communication handovers which could, over time, degrade services.

3. **Engage with remaining key partners and vendors to re-negotiate contracts and expenses.** You can sometimes reduce costs by reviewing the commitment of your existing relationship. With more predictability than you may have seen in Grow, you might be able to increase commitments to some suppliers or consolidate services. Be careful to balance risk if you are considering working with only a fewer suppliers.

4. **Outsource non-core capabilities to specialist service providers.** Where you have repeatable processes, consider outsourcing non-core capabilities to cheaper suppliers. Business Process as a Service (*BPaaS*) can offer models which are more

---

[57] David J. Anderson & Associates (2010). *The principles of the Kanban method.* Available at http://www.djaa.com/principles-kanban-method-0

cost-effective as can other Platform as a Service (PaaS) offerings. System maintenance, periodic services and delivery bursts or other activities which are not core to your own business, might be better served by a partner providing these as a service.

5. **Reduce non-value-added tasks and expenses.** Consider stopping travel not related to sales and placing limits on expenses internally or through existing partnership agreements. However, be careful to consider the impact on morale and indirect productivity effects on performance before making these decisions. It's easy to see costs in a spreadsheet and make decisions, but less obvious are the secondary effects of these decisions.

6. **Reduce team size and rebalance skills for your business needs.** Repeatable processes should be automated where possible as this could provide considerable savings. Human potential can sometimes be wasted if limited on repetitive tasks. People in your organisation could be more valuable if they work on products requiring exploration and discovery. We have often observed situations where companies working on new ideas and initiatives bypass using existing skills in the organisations and seek new hires. This adds costs to recruitment, training and increases knowledge drain.

7. **Observe and improve quality where rework is a cost.** Where can you reduce defects, waste and time invested? Remember the costs of reducing defects should be less than the savings made.

Do whatever it takes to streamline your process, eliminate inefficiencies and focus on the bottom line while not compromising customer value. For example, as you will no longer be adding new features or limiting the updates to your product, it seems sensible to reduce your new product development spend. However, you may want to improve customer services, logistics and other operational functions. If your product needs a sales force to deliver revenue, you may not want to cut that.

You can also consider reducing the features of your product. Look at the actual customer behaviour against the features of your product to identify where usage is light or non-existent and analyse the impacts of removing that feature. It's not uncommon to generate a lot of waste as a result of feature creep over time, which not only results in increased cost to maintain, but also reduces value from your core product and can disappoint customers with complexity. If part of value delivery is a great website, make sure that you are fixing all bugs as they arise.

6.3 Key strategic partnerships

As we have already noted, selecting and reviewing key partners should be an ongoing activity in your business to refine your business model. You will likely already have in place key partnerships which have been crucial to make your business grow and successfully reach Sustain. However, your business model is constantly under pressure from environmental changes, as are your partners. Therefore, like other components of the business model, your key partners should be actively reviewed to ensure they are still the best choice for your business.

When reviewing your partner selection and needs, you will need to have a structured performance-based relationship where quality and reliability of the service are understood by both parties and managed. It's important to also factor in the risk of the partnership to both businesses. A key partner will also need to see the benefit and potential of the agreement. Exploiting partners too much can be a risky business, particularly for key services in complex environments.

When negotiating a relationship with a key partner, you should focus beyond the initial transactional value of the agreement. Where both parties can have beneficial outcomes, you are often more likely to have a better relationship beyond the point of transaction. This is a key aspect to continuous improvement which can go a long way to optimising performance. Remember as your business grows and responds to change, you want key partners who can respond to that change with you.

Nine steps to establishing key partnerships in Sustain

1. **Define your needs and goals –** By the time you get to Sustain, you should have significant success and knowledge around your business model. You will want to ensure partners align to your needs and vice versa. Your partner selection is not just based on your journey to date, but supporting you on where you need to go. This is why you need to clearly define your needs and goals before engaging with potential partners.

2. **Achieving a win–win outcome** – Selecting a partner is a two-way process and when selecting partners consider what is in it for them as well. Why do they need to partner with you? What capabilities can they offer? Will they be incentivised to work with your organisation and invest in change as you evolve? Having a partnership where both parties benefit is more likely to yield a closer and more fruitful relationship, particularly when considering support, priorities and adaptive processes.

3. **Aligning on quality –** Many suppliers and partners can deliver contractual quantity. However, the cost of defects will impact your business if quality isn't assured with low variance on expectations. Explicitly agreeing and aligning on quality expectations through your process ranging from raw materials to customer service expectation, can significantly reduce cost. It's beneficial to share to agree on measures of quality and align on the KPIs that matter through regular collaborative reviews.

4. **Skills and complexity of service** – If you're seeking specialist skills which are highly complex in nature, it's likely that your choice of partners will narrow. In some cases, this complexity puts extra emphasis on the win-win outcomes defined above. As a general rule, the more complex the service we are seeking, the closer it should be to the business. It is not advised to make the acquisition of feedback and knowledge more complex than it needs to be.

5. **Beware of the hidden costs of outsourcing** – Making the big decision to outsource based purely on like-for-like operating expenses is naive. When choosing a partner to manage a key service in your business, you need to consider hidden costs which are often overlooked. When working with internal resources, you often have value beyond what's captured purely in the defined process, including an ability to change without the restriction of contracts in place. When these relationships are not understood you can end up with a restrictive partnership, which make you less able to respond to change; hindering your ability to improve as a business.[58]

6. **Due diligence –** With a product in Sustain, you have a lot to lose if you don't select the right partner. Before making decisions, define what you need to know and identify the risks. Depending on your domain, there may be restrictions in place you need to adhere to regarding regulations, transfer of data and SLAs. Seek the right council and ensure you select a partner who can support your cultural and operational demands with explicit experience operating within the markets you operate.

---

[58] Forbes / Rubin, J. (2013). *The hidden costs of outsourcing.* Available at http://www.forbes.com/sites/forbesinsights/2013/03/29/the-hidden-costs-of-outsourcing/

7. **Transparent measurements –** It is important to have a clear understanding of what we are expecting from the partnership on an ongoing basis. We need clear metrics of success agreed upfront. It's advisable to capture both direct deliverables and peripheral components which should be measured by outcomes. For example, lead time is a direct deliverable, whereas customer satisfaction would be an outcome with many contributing inputs.

8. **Burst capacity and seasonality –** Testing your partner's ability to respond quickly to change is a key component to keep costs low. When you have spikes in demand from seasonality, your entire network needs to respond. Failure to do this may severely impact your business in lost revenue, damage to brand and future opportunity costs. It's advisable to ensure your partner provides customer references around their ability to support seasonality.

9. **Trial and evaluate –** With so much at stake for your business, we highly recommend that you undergo a trial or pilot programme before you take partners on. You should also consider running pilots before you bring existing partnerships to an end. A trial programme will allow for the discovery unknowns which could not be otherwise foreseen.

🔍 Tip

If you have decided to undertake a re-organisation of your company, beware of assumptions from afar. Often these sensitive matters are handled by third-party consultants, agencies with sometimes junior consultants who have little or limited knowledge of your company culture and ways of working. Looking at a job title on a spreadsheet doesn't always align or explain what a person or team may actually do or the value they provide. This is prolific in large enterprises where job titles are the same across business units, but each operate differently.

Embracing a Genchi Gembutsu[59] mindset would be recommended before making centralised or distant decisions. Although this can sometimes focus on process improvement, the notion of going to see, observe and understand the environment you are looking to change is highly recommended. You may find areas to improve as well new information to inform your decision.

---

[59] Economist (2009). *Genchi Genbutsu.* Available at http://www.economist.com/node/14299017.

6.4 Platform innovation and sustainability

It might be easy to infer that products in Sustain minimise further product development due to the emphasis on operational efficiency. This is not necessarily the ase. It is likely that digital products will require a higher proportion of on-going investment support and incremental innovation. Given that digital products sit on platforms and operating systems which are constantly evolving, you will need to keep pace to keep your customers and dependent platform lifecycles. Just looking at the iOS lifecycle alone, there have been nine major releases from 2010 to 2017.[60] The rapid adoption by loyal users also typically shows that approximately 70 per cent of users will upgrade in just over a month.[61] If you don't manage and prepare for this switch, you risk losing your customer base in a matter of just a couple of months.

Almost all products that contain software are likely to need of ongoing investment and maintenance, particularly as customers demand more of the products. Consider that a modern high-end car actually contains more lines of code than Facebook or Microsoft Office.[62] Also if you recall from earlier sections of the book, your product is not just the object sold, but it is part of your business model. As such, it is important to consider how you will maintain all systems and platforms which sit across each element of the business model.

If your product is physical, your channels and customer relationships may be digital. If you sell books today, support for this product may require keeping aligned with Amazon's integration needs for distribution. You may have a website which needs to be accessible to evolving mobile platforms. You may also have to manage orders and customer support through evolving channels such as Twitter, Facebook and more.

Keeping pace with platform innovation requires continuous observation and management to manage your product in its lifecycle. When Microsoft update the Windows operating system, there are subtle changes which need to be considered. Operability, performance and customer interaction will impact your product and you

---

[60] Wikipedia (2017). *iOS Version History*. Available at https://en.wikipedia.org/wiki/IOS_version_history

[61] Mixpanel Trends. *How quickly are users updating to iOS10*. Available at https://mixpanel.com/trends/#report/ios_10

[62] McCandless, D. (2013). *Million lines of code - codebases*. Available at http://www.informationisbeautiful.net/visualizations/million-lines-of-code/

should be working with the vendors and beta developer programs to align the customer migration path.

To highlight this point, consider the platforms supported by Netflix. It could be argued that Netflix is in Sustain in the USA and Grow for other regions, and has to keep pace with platform innovation.[63] Netflix is actively improving operational excellence on profits; a key activity of Sustain. Being available on smart TVs, mobile devices, PCs, streaming media players, set top boxes and games consoles is no small undertaking. As the market gets more saturated by device offerings, reaching and maintaining customers is an ongoing challenge.

For each of these customer streaming device categories, there are a range of systems and manufacturers to support. Netflix customers can have any combination of each of these and expect a continuous experience across platforms. As each of these platforms evolves and new platforms are introduced, Netflix has to move with the customers' upgrade patterns to Sustain their business. Therefore, platform innovation is a core activity for Netflix to maintain their market share at a minimum. Failure to do so will result in a decline in customer retention.

When reviewing your product on a regular basis during Sustain, you should be observing and monitoring the effectiveness of existing channels. Detecting weak signals of change in the market and technology platforms could ensure that you are riding the wave of change and staying ahead of the curve. Do not assume that Sustain is static stage in the Lean PLC that is all about just cutting costs. Nowadays, to sustain customers, revenues and profits, you have to keep innovating. The only difference in Sustain is the strategic focus of the innovation.

6.5 Actively managing products in Sustain

As with Grow, the hope is that we can keep our product in Sustain for as long as possible. As such, unlike the left side of the Lean PLC, products in Sustain are managed via a regular review process. We advocate using regular data-driven reports and

---

[63] Forbes (2017). *Netflix subscriber growth continues unabated, as margins improve.* Available at http://www.forbes.com/sites/greatspeculations/2017/01/19/netflix-subscriber-growth-continues-unabated-as-margins-improve/

scorecards. Having a regular cadence of review whether it be monthly or quarterly, will help you structure key data points and observe key trends. This allows you to respond with the right corrective measures and exploit emerging opportunities. Remember that the key activities in Sustain include:

1. maintaining market share, customer numbers, revenue and profits
2. keeping customers happy by making sure our product consistently meets their needs
3. optimising operation by reducing costs and improving efficiencies
4. maintaining sustainable relationships with our key partners
5. tracking evolving technology platform and ensuring that our product stay relevant and aligned.

Each of these points should have associated metrics on your dashboards which can be informed by your business systems and activities. Develop internal capabilities to report on these key points regularly, overlaid by your Sustain hypotheses. This will have a profound effect on your ability to manage your business and respond to change. Below is a Sustain template you can use, but feel free to extend and contextualise the sections and data points to best fit your company.

6.6 Prepare for Retire – nothing lasts forever

It is an inevitable part of the PLC that a product or service will begin to decline. At some point, there will be significant declines in customer numbers, revenues and profits. Changes in technology and customer preferences can render our product obsolete. Within the Lean PLC, we recommend active portfolio management which requires us to make decision about when to retire our products. This empowers the organisation by allowing it to systematically manage the retirement of its own products. This is different from having zombie products within the company or being surprised by a sudden loss of profitability.

There are several factors that can influence the decision to retire a product.

☐ Three of more quarters in which financial losses have been made.

☐ The technology landscape has moved on in a manner that renders our product obsolete.

- The product no longer delivers values to customers or meets customers' changing expectations.

- The cost of keeping the product going is much greater than the revenue it is generating. This state violates a key principle for remaining in Sustain.

- We have been disrupted by a startup with a better product and business model. Even as we fight back, we might also consider retiring the product.

- We have successfully disrupted ourselves. We have created a better product with a better business model. We should then actively sunset our own product, rather than just keep it going.

- There has been a shift in our strategy, and the product is no longer a fit.

As part of the regular review of our products during Sustain, the question of whether to retire the product should be consistently raised. When we make the decision to retire, we should base it on evidence and validated learning. We should not be making decisions based on conjecture. Our conversations should explore the cost of both keeping and retiring the product, impacts on customers and the option to move customers to our other products and services. We should also consider what other residual value we could glean from the product. Retiring a product is not the only option; the product can also be sold to other companies. In the end we need to have a clear exit plan that assesses financial impacts as well as any additional resources that may be needed for the retirement process.

Sustain to Retire SUBMISSION

Product ownership

Investment board	
Business sponsor	
Product owner	

Product overview

Product name	
Idea description	
Strategic fit	

Product performance

Please describe the financial performance of the product to date.

	Revenue	Costs	Profit
Forecast			
Actual (or re-forecast based on actuals			

Sustain lessons learned

Provide an overview of the sustain hypotheses that you have tested since the last review and the lessons you have learned.

We believe that:	
To verify that we:	
And we measure:	
We learned that:	

Unexpected learnings

Provide a summary of the unexpected learnings you got during Grow.

Planned activity

What are key tasks and milestones involved in retiring the product?

Financial projections

Since you are asking to move to Sustain, please complete the financial data below.

Expected revenue over 3 years	Expected costs over 3 years	Profit margins over 3 years (%)

Resources and funding requested

To Sustain the product, we are asking for (e.g. **dollars, time, people**):

* Please note that you also have the option to remain in Grow, go back to Validate or move straight to Retire. If either of these options are what you are choosing, then you need to adapt this template so that you can update the product council on work done so far, key lessons learned, what you plan to do next and the resources you need.

A case study – Paradox Beta

Paradox Beta is a company that focuses on developing niche software for a specific industry domain. With a mature and well-established customer base of leading brands, Paradox Beta's cash cow product was in Sustain. After many strategic initiatives, the business observed that the customer demands were changing to cloud-based services. This observation resulted in a rash decision that the company needed to exploit the existing product in Sustain to protect revenue and grow the business. The profits from the product in Sustain were invested into a new cloud-based version of the product.

Paradox Beta simply created the prototype without doing any business model exploration or validation. Interesting, the product gained rapid traction from the technical success of the prototype. As Paradox Beta started to operate in this space, new complexities emerged which were not present in the previous business model. One could say that the business shifted from a *complicated* and *simple* domain to a *complex* domain unknowingly. The company continued to make public commitments and increase investment. This added pressure to internal teams.

During this business model change, the previous partnerships were not structured to support the new business model. Interestingly, these partners were not part of the initial stages of this journey. The strain imposed by the business on the existing partnerships resulted in stretched resourcing and tensions within long-standing agreements. In some cases the partners significantly increased prices, which rendered the previous business model less profitable, or they declined to continue services. Adding to these woes was the negative impact on customers. Customers who had been promised the same level of service they had become accustomed to were now moving away; decreasing customer retention. These laggards had been unwittingly turned into early adopters and they definitely weren't happy with that.

During the panic and emotive decision making that was taking place, the company also selected and bet upon an embedded new technology through a limited partnership in their solution. Without strategic analysis, they did not qualify the strategic partner or establish the channels of support which were not offered by the partner. The cost of this meant that Paradox Beta had to invest heavily in domain specialists and services to support the new technology, which again took further investment and resulted in further losses.

The partner then took a different direction, which resulted in a discontinued relationship. Paradox Beta had long-term contracts in place with customers and brand reputation at stake. They eventually had to swap out the technology at additional cost and time. The outcome in the end resulted in the business having a less profitable business model for their product in Sustain, deprecating the new product, losing key employees and ultimately wasting significant time and resources that could have created new opportunities.

There are many lessons observed in this case study, which are of course easier to observe in hindsight. No doubt the leadership at the time of making the decisions made them with hope and good intent. However, had they applied the key stages of Idea through Validate, they may have learned more about the complexities and effort needed to develop a new business model. During this process they may have also evaluated and qualified their partners in a very different way at all levels and may have limited the negative impact on their existing customer base.

*Please note that the name Paradox Beta was used to protect the name of the real organisation that was involved in this work. However, the case study and the lessons learned are real.

7

Retire

7.1 Welcome to Retire

Unfortunately, all good things must come to end. When a product comes to the end of its life, there is often an emotional sense of failure or loss, sometimes with and added fear of change. As such, there is often a reluctance to retire products unless problems are severe, obvious or inherited; a costly mistake for innovation. If you wait until you have to, retiring a product might be costly in terms of both finances and brand equity.

Actively looking for products that are on a downward trajectory, have limited potential or are no longer needed to deliver your current strategy, is a healthy endeavour for a company's long-term growth. Growth and innovation requires continuous investment of resources and energy. Whatever size your organisation, you will have a finite capacity to manage products at every stage of the lifecycle. A mindset of minimising waste and maximising value through applied effort will undoubtedly improve your product portfolio.

The opportunity cost incurred by dragging out products that should be retired is wasteful, high risk and to some degree negligent. It's incumbent upon decision makers to ensure their efforts are applied to the most rewarding endeavours to deliver on strategic goals. This is not just for the benefit of capital return for investors, but for the security of the company's future and those that depend on it.

Recycling retirement capital to the top of the innovation funnel or doubling down on other promising endeavours should be strategically encouraged to improve the chances of long term corporate survival. Another benefit of actively considering retirement candidates is that the activity can reveal new ways to utilise and apply existing capabilities and technologies. There are ten key points to consider when retiring products. We outline these below.

7.2 Ten key points to consider when retiring products

As you make your decisions to retire, you will be considering financial performance, customer feedback and other trend data. However, these data alone are not enough to make a conclusive decision to retire a product. Beyond finances and customer feedback, we need to make other considerations in order to make informed decisions, manage customer expectation and support employees that are affected by decisions to retire. We also have to consider the risk and impact of our decision. We have identified ten key points that can help you make your decision and manage the process of retiring a product.

1. Have you identified any dependent service level agreements and contractual obligations?

The you have decided to retire is likely to be part of an ecosystem with contractual service level agreements (SLAs) in place. Failing to consider these SLAs could result in legal repercussions or brand damage which could far exceed the cost of keeping the product alive. Look to identify these SLAs first before making any decisions and consider the cost of breaking the contracts, the impact on your brand and other strategic opportunities. This is particularly important if there is a chance of negatively impacting your customer or partners.

Consider pharmaceutical companies which may be contractually obliged to maintain reserve stocks for a period of time. Customers who utilise a product as a critical service may have contracts with the company to protect this service. In some companies, minimum notice periods may be specified and in service companies there may be obligations to support customers for a specified period of time. You may also have contractual obligations to hand over the intellectual property or source code if you are a software company. All these things have to be considered when making the decision to retire a product.

2. Do you have sufficient customer service and marketing resources in place to manage the additional support required to support product retirement?

Following an announcement to retire a product, you will likely see an additional spike in customer support activity. A lot of customers will have concerns about the impact of your decision to retire a product. As soon as there is an announcement, customers will seek more information and reassurance. Larger companies will also need to consider market opinion which creates a media interest that needs to be managed.

You may need to consider training and investing in more customer services for the period of the retirement, even though you might be reducing other operational activities. Other costs and impacts should be considered in marketing to handle PR issues, announcements and communication campaigns, direct contact with customers and other such activities. Retire is a phase and not a transactional decision. As such, this will often require project-related activities, diligence and managed oversight.

Consider the impact of the retirement decision on your internal teams as well. Announcing products for retirement can have an adverse effect on your staff across the company, beyond the directly affected product team. Internal messaging and support should be managed internally to assure those who may feel insecure as a result of the decision. Not handling the internal impact could result in talent loss and may damage the company's reputation as a desirable place to work.

3. Do you have sufficient technical capabilities in place to manage the retirement phase?

Just as you need to invest in customer services and marketing, due diligence should be undertaken to ensure that we have the technical and specialist skills needed to manage the retirement phase. When retiring a product, you may need to assign and make available dedicated teams, persons or need to recruit new skills in order to handle the activities of the phase.

Consider for instance data management as a retirement activity. You will likely need to manage the destruction and/or the migration of data securely; and provide evidence of standards being met. This might be part of your process for reducing the impact on customers, allowing them to migrate data or assure them with evidence of destruction of their data. If you are ISO 27001 compliant you will already be aware of the standards expected, but you might need to increase activity to manage these processes. As such, it is critical to ensure that we have these capabilities in place before we retire a product.

4. Have you identified dependencies and communicated the retirement intention and plans to all stakeholders and business areas?

Investigating the impact on dependent systems is an exercise well worth undertaking. Many dependent systems can be easily overlooked where features sitting in the corner somewhere could be unknowingly providing valuable contributions to other products and services within your company's portfolio.

An exercise analysing the dependencies on a product can reveal unknown risks from dependencies that may be in place. This exercise allows us to understand the ecosystem impacted by the product that is marked for retirement. The exercise can also reveal components of intellectual property that can be extracted and further utilised. If there are dependent systems which utilise the value of the service, ensure that the cost and revenue is proportionately attributed to better inform your retirement decision.

It is also important to engage with key functions and stakeholders before making decisions. This may reveal insights which would otherwise be missed by the using data alone to make decisions. For instance, sales teams who might be in front of the customers everyday may have empirical knowledge as to what the customers may migrate to as an alternative or how they may react. This could affect your tactical decisions and should be considered for improving outcomes.

5. Have you identified how assets, knowledge and resources can be utilised by other areas of your business or partnership network?

Consider the retirement of products as a learning opportunity. Before closing the doors on a product or service, consider what could be learned and utilised. You are likely to have very experienced staff, which could add considerable value elsewhere in the business. A declining product should not be presumed to be associated with a bad team. Products failed for many reasons and the team could be very valuable when applied in other areas.

A lot of company cultural knowledge resides in your employees which would take considerable time and effort to retrain if you were to lose the team. The people associated with a retiring product should not be associated to any notion of failure. Making such associations could be an expensive mistake. For example, in operations you may have in-depth knowledge of customer usage data, which other products could use and benefit from when innovating.

The product being retired may also have key assets that could be of ongoing value to the company. These assets should not be discarded as part of the retirement process. They can be added to current products, used as a basis for creating new products, preserved as intellectual property for future use, sold to other companies or licenced for revenues.

> ### Q Tip
>
> Consider positioning the employees of the retired product into an internal marketplace for talent acquisition. You can help employees utilise their existing networks, update their recognised, CVs and even hold job fairs. If you're an enterprise with many products, you may have products moving into different lifecycle stages who are recruiting. This can help employees move to new and interesting opportunities.

6. Have you provided options for customers to export and manage their data and other invested assets?

The customers of the product you are retiring, maybe your customers of tomorrow. Treating them well and considering their needs when retiring products is in your interests. There are many things to consider around how the retirement process will impact your customers and this should be at the forefront of your decision making.

One of those concerns is customer data. Customers will have concerns about how you will manage their data as part of the retirement process. Being transparent about this could potentially settle their minds and reassure them. This can be done by explaining how their data will be destroyed and managed as mentioned above, but you could also provide customer with an opportunity to extract their data and for potential use elsewhere.

Consider for instance if your product is a wearable technology that tracks customer health data. If you were closing down this product, customers may want to download their data in a universal format to utilise on an alternative product. If you are looking to cross-sell an alternative product, you may want to migrate the data to this other product as part of the process. Alternatively, if you are leaving this product space, you could explore ways to format or export the data for the customers to migrate to other platforms.

Supporting the customers and/or partners to self-manage any invested resources they have in your product will require analysis, consultation and effort. Some of the skills you may have for innovating and exploring customer needs to create a product, could be directly applied in understanding their needs for retirement. Getting this right will help protect your brand and reputation, as well as reduce any negative impact that may otherwise be felt if not managed well.

7. How will you manage the customer acquisition pipeline?

When a product is selected for retirement, it's important to close the pipeline of new customer acquisitions. This may include the capability of signing up for the product or creating an account on a digital platform. You should extend this to discovering the product through various channels and informing sales and marketing so that no future sales will be negotiated.

You may wish to stagger the execution of the retirement phase with different functions, events and departments. For instance, your sales force might need to be updated, trained and encouraged not to sell the product before an official public announcement and be in position to answer customer questions. If you have consulted such functions as part of your due-diligence process, this should not come as a surprise to them.

8. Have all finance issues been identified and resolved?

Retiring a product is not free. You may be stopping other key functions, but there may still be costs to retiring the product. All these retirement costs should be identified and presented as part of the cost of retirement. It is expected when executing the retirement process of a product that additional finances will be required to manage the points identified in the process. In addition to this, any products costs and loss of revenue should be identified and resolved using financial forecasts and budgeting so that capital can be managed for future investment.

9. Have alternative products been identified to help manage and transition customers?

Just because we are retiring a product does not mean that we should lose all the customers connected to that product. Instead would should offer customers the

opportunity to move to an alternative product where possible. Transition opportunities should be identified and migration expectations should inform our retirement strategy. We should work to identify customer migration paths and create processes to help make this transition as painless as possible for our customers.

10. Have you explored the option of renewing and refreshing the declining product?

A final question to consider before we pull the trigger is whether the product can be renewed and refreshed with a different business model. It is very easy for products with repetitive operational cycles to decline, and this can result in opportunities being missed. Market conditions can change, new channels can open or technology can advance. As such, it is key to explore whether these changing circumstances provide an opportunity to have a second growth curve on our product's lifecycle.

In this regard, we can go back to ideation and run brainstorming workshops. We can redesign our business model by considering how to make it more adaptive to the new business environment. This work would take the product back to the earlier stages of Idea, Explore and Validate; as we test our newly designed business model before we scale. The potential outcome could be new innovations which have been trapped inside the business unit or product team.

7.3 Refreshing your product with adjacent innovation

As already noted, retiring a product is not just about releasing capital or reducing waste, but can be a proactive process to discover more valuable applications for the declining products or their components. By seeking more value, you may uncover new and innovative business models which would not otherwise be discovered. Remember that it is highly likely that the business environment has considerably changed since the product's inception and the business model could be revisited.

A powerful example of the reapplication of a technology is well described by Naji Radjou and Jaideep Prabhu in *Frugal Innovation – How to do better with less.*[64]

The Retire stage in the Lean PLC can encourage a team to revisit the breadth of new opportunities that might now be attractive to them but were priced out at the time of

---

[64] Radjou, N. and Prabhu, J. (2015). *Frugal Innovation: How to do more with less.* London: The Economist.

original product conception. Technologies which have aged in wealthier western societies, could have a second life in less affluent countries where cost was originally a significant barrier. New technologies when first conceived and supported by early adopters are usually expensive to scale and thus constrained for wider market application. With a reduction in cost, the potential adjacent markets could be significantly broader. In some cases, these technologies can end up creating new categories in new markets.

To put this in context consider 3D printing. The entry level for 3D printing is still a barrier for market adoption for most. Over time as the price drops, new markets and dependent economies may emerge. The intellectual property (IP) within the older devices is in some cases recycled into newer products with cheaper and more reliable materials, making products more accessible to a broader audience. As this happens, the technology can become transformational through adjacent innovation opportunities and broader market adoption.

A case in point is the work of OpenBionics.com who use 3D printing to support amputees. Traditional prosthetics hands can cost up to £90k and take months to fit, making them too expensive for most. Using the latest 3D printing and body scanning technologies, Open Bionics can create bionic hands for just £2k and can take just five days to fit. As such, if you were in the production of original prosthetic limbs, could you revisit your business model to utilise 3D printing as above? By doing so you could lower price and increase your total addressable market.

This approach may seem like cannibalisation of your business model. However, if the product is already in decline, this may be a great option for your company. It's important to actively measure how each product is performing against your strategic goals. Whatever size company you belong to, you will have a finite capacity of idea generation and execution. Releasing the untapped potential from existing products or teams by actively seeking to retire products, can help propel you forward and innovate.

SUBMISSION TEMPLATE

Retire review SUBMISSION

Product ownership

Investment board	
Business sponsor	
Product owner	

Product overview

Product name	
Idea description	
Strategic fit	

Why Retire?

What is your rationale for retiring the product?.

What impact is this having on our customers and key stakeholders?

What are your plans to mitigate impact?

What are key tasks and milestones involved in retiring the product?

Planned activity

What are key tasks and milestones involved in retiring the product?

Financial impact

What will be the financial impact of Retiring the product on our company?

Expected revenue over 3 years	Expected costs over 3 years	Profit margins over 3 years (%)

Resources and funding requested

To complete the Validate stage, we are asking for (e.g. **dollars, time, people**):

A case study – Nintendo

Following a series of successes in the early 1980s with arcade games, Nintendo made plans to create a cartridge-based console. Nintendo president Hiroshi Yamauchi firmly stood behind his vision of providing a family computer (famicom). His vision focused on the simplicity of the cartridge and controller, allowing the product to be more accessible to a wider range of family members. This was a key factor and resulted in having no needed for a keyboard, which at that time many found intimidating.

This idea eventually led to the creation and release of the NES, which was launched on 15 July 1983. Despite there being many games consoles prior to this idea, including the Binatone TV Master, the Atari 2600 and Nintendo's own Colour TV-Game series, the NES leapt forward with games which engaged families across the globe.

The console offered many innovative features at the time, especially the subtle introduction of the D-Pad (Direction Pad), giving gamers more control and user engagement. In addition to this, Nintendo created some of the most iconic games of its time ranging from Super Mario, The Legend of Zelda, Donkey Kong to name a few. This grew Nintendo's foothold as a category leader. The games have long outlived and evolved beyond the original NES capabilities and today sit across many platforms including our smartphones. However, without any doubt the NES was a huge defining moment for Nintendo.

Now younger readers or those less familiar with gaming history might see the NES today as an antique of sorts. This is especially true in the advent of Augmented and Virtual Reality with incredibly rich and powerful consoles and smartphones. However, that doesn't mean there is no value left within the NES ecosystem.

Thirty-three years on in 2016 Nintendo decided to resurrect the old NES and package it into a smaller modified version, preloaded with 30 games called the Classic NES. Despite the obvious advancement in technology in the last 33 years, there was significant demand for the Classic NES, way greater than Nintendo even expected. Nintendo anticipated this would just be an additive product for a seasonal offering, but demand was much higher.

In comparison the original NES sold approximately 2.3 million units and the Classic NES sold 1.5 million units following increased production in just a matter of months.[65] Despite the demand, Nintendo did eventually stop production to most people's surprise. Their reasoning was that the NES was not aligned with their strategic interests. However, the demand for the Classic NES has since given rise to the resurrection of other older games consoles including the Commodore 64 Mini,[66] Nintendo SNES Classic, Ataribox and more.

This example with Nintendo demonstrates two main things. First, it is possible to successfully retire your own products, before the market kicks you out. Second, there may be trapped value in older products and platforms you offer. Just because something has been around for a long time, it doesn't mean there is no market for it. Clearly Nintendo revisited a core product decades later, applied some innovative thought into how it applies to the modern context and re-invented the technology which could have otherwise been easily overlooked.

[65] Time (2017). *Nintendo says it sold over 2 million NES classics.* Available at http://time.com/4759594/nes-classic-millions-sales/.
[66] Swapna Krishna, E.S. (2017). *A mini version of the Commodore 64 is coming in 2018.* Available at https://www.engadget.com/2017/09/29/commodore-64-mini-releases-in-2018/

8

Start Tomorrow

8.1 Start tomorrow

We began this book by remarking how the world is changing rapidly. Corporate leaders and startup founders are all feeling the pressure to keep up. Traditional approaches to management are clearly no longer sufficient to take our companies into the future. Training our product teams to use lean and agile methods is a great step forward. But this training on its own is not enough. We also need to transform how our companies develop strategy, manage and invest in products. So why wait to begin the change? Let's start tomorrow.

The Lean PLC as a model was designed to provide a new management toolbox that connects product development best practice with the right investment management tools. Doing the right things at the right time – asking the right question at the right time. Bringing these practices into your organisation is not an easy task. Every company has antibodies that resist change. There will be political inertia and active resistance to transforming how people work.

Such change will not happen by itself. You will need to create a small team whose job is to champion the Lean PLC within your organisation. In our experience, this need not be a particular role or job title. It tends to be people who are passionate about your company's future and ways of working. You might have a head start if you already have lean and agile coaches within your company. However, nothing beats a team that understands lean thinking and is also passionate about sharing their knowledge and helping other people continuously improve.

In this chapter, we will share some of the lessons we have learned implementing the Lean PLC in established companies. We will outline the following eight steps.

1. *Understand the context*: Focus on why you are transforming your company and begin with discovery.

2. *Get leadership support*: Work with both C-level executives and middle management to get air-cover and support for the Lean PLC.

3. *Work with early adopters*: Transformation does not happen as one big bang. Work with early adopters to develop a bespoke model for your company and get an early win.

4. *Work with enabling functions*: The Lean PLC does not work without support from enabling functions such as finance, technology, operations and sales.

5. *Map your products*: As the framework begins to take hold it is time to map all products in your portfolio to the Lean PLC stages. This is a great way to get a portfolio view of your products and socialise the framework.

6. *Create playbooks and tools*: A playbook is a great way to socialise the Lean PLC within your company. You will also need investment decision-making tools. Such practical tools will help teams to bring the Lean PLC to life in their day-to-day work.

7. *Establish investment boards*: Now that you have a framework for doing the right things at the right time, it is time to set up an investment board to make investment decisions based on the Lean PLC.

8. *Build communities of practice*: Communities of practice is where all of this comes together. The goal is to transform the culture of our organisation. Active and engaged communities are a great way to build a culture of experimentation and innovation.

Understanding context

When working to transform a company, it is important to be knowledgeable and understand what you are talking about. If you are not knowledgeable, the challenges of transformation will eat you alive. So by reading this book and other writings on this topic, you are already taking important steps. Some experience of doing the actual work also matters. And the one thing we have learned from our own work is that one-size does not fit all. We believe that the Lean PLC is a great framework of tools and practices. However, we also believe that the framework has to be adapted to fit the context of the company that will be using it.

Our case studies also show that there are some great companies that are applying Lean PLC principles really well. However, you must resist the allure of implementing what has worked in other companies within your own organisation. Most standard frameworks and practices from other companies will break when they face the reality and complexity of your own company.

This is why we believe that all Lean PLC implementation work must begin with discovery. Do not just take the standard framework and apply it. First, do the hard work of understanding how your organisation really works.

- ☐ What are the unique challenges within your company?

- ☐ What has worked successfully before?

- ☐ What initiatives have failed and why did they fail?

- ☐ Where are the roadblocks and resistance to lean thinking?

Such discovery work will also uncover key allies and champions that you will need as your transformation work begins.

Part of understanding context is also understanding the reasons why the company wants to adopt the Lean PLC. Before we begin the transformation, there has to be organisational clarity as to why the changes in ways of working are necessary at all. As such, it falls on the leadership within your organisation to identify the reasons for the adopting the Lean PLC. Leadership has to be clear about the how the world is changing, key trends that are affecting their business and how they plan to use the Lean PLC to respond those changes. This also means that the Lean PLC cannot be successfully implemented in any company without active support from executives and leadership.

Get leadership support

This chapter assumes that you are adopting the Lean PLC as a way to transform how your company works. However, such transformation is not always necessary for teams to begin using the Lean PLC. People can just start using the Lean PLC within their own teams to develop products in a new way. If this is your focus, then you have to be clear about the choice you are making. Be prepared for some clashes with other teams and leadership within your company that are not using the Lean PLC. You will have to be politically savvy in order to protect your team and allow them to keep using the Lean PLC.

Our own experience has shown us that teams eventually have to change some elements of how their company is run in order to fully benefit from the Lean PLC. We have also learned that such changes will not happen without leadership support. Without executive buy-in any efforts at transformation will be dead on arrival. A company's culture is determined by what its leaders reward, celebrate and punish. Guerrilla movements that try to change a company's culture via the grassroots almost never work until a key executive is convinced of the value of the change and becomes its champion.

As such, we would recommend that you identify the main players within your company that you will need support from. This often includes ensuring that you have C-level support for the Lean PLC. Such 'aircover' will be helpful in removing the inevitable roadblocks and resistance you will face. It is particularly important to identify a C-level champion who will actively support your work and act as a diplomat to help you get resources.

Although it is absolutely essential, C-level support is not enough. While executives may set the overall strategic direction of the company, the people that actually run the company day-to-day are the middle management. Innovators often refer to middle managers as 'permafrost'; the place where all innovative initiatives go to die. But this is a mistake. Middle managers are not simply ignorant MBAs who are stuck in the past. They often have important goals to achieve that have been set for them by the C-level executives. Their bonuses and promotions depend on delivering on these goals. As such, we have to work with them and show them how the Lean PLC can help them achieve these goals.

Some teams try to bypass middle management by going straight to C-level executives. But even if you get C-level support, you are still going to need some help from middle management to implement any innovation project. While they may not actively fight you, middle managers can passively resist your work by using procedural delays that will eventually kill your project. In fact, the investment decisions that are a key part of the Lean PLC will often need to be made by middle managers. As such, the Lean PLC will not work without allies within the ranks of the middle management; make sure you have their support.

Work with early adopters

One mistake that energetic innovators tend to make is to try and change their companies in one big bang. So they spend a lot of time advocating for change. They find executives that will listen to them and they preach about how the world is changing and that the company needs to change as well. The problem here is that we are just talking. We are not doing the hard work of showing leadership how useful the Lean PLC really is. This makes it easy for them to dismiss us as crazy mavericks, often using less flattering language!

We do not recommend this approach. We bear the scars of having tried it one too many times. Preaching and pontificating about innovation encourage resistance from the organisation. What you need is an early win that shows the company how the Lean PLC really works. This means that you have to work with early adopters. Within every company there will be pockets of leaders that get it. They understand how the world is changing and what needs to be done. Working with these leaders first means that our transformation agenda will not face too much resistance while it is still in its nascent stages.

We should select a high-profile and important project within our early adopter's business unit and work quickly to demonstrate that lean PLC methods work really well. We should document this early win with data and begin celebrating it widely within the company. Such an early win buys us goodwill within the organisation. In fact, it is the perfect tool for getting the leadership support we need to implement our transformation programme. As we succeed with early adopters, other business leaders will be attracted to our transformation agenda.

Working with early adopters also allows us to learn more about our company and develop a bespoke Lean PLC framework. Working on an actual project will reveal pockets of resistance and other challenges that we did not anticipate. We will also learn about things that work really well within our company. Since we are working in a limited space with a small group of early adopters, we can use any lessons we learn to adapt and change our model. This approach ensures that we have a model that is likely to work when we roll it out across the rest of the company.

Work with enabling functions

Our work to understand the context, get leadership support and an early win is useful for developing a bespoke Lean PLC framework for our organisation. However, the tools and practices of the Lean PLC will not work without support from the enabling functions within the company (e.g. HR, Legal, Finance). Their role in designing how the framework will work within your company is crucial. Enabling functions also play a key role in the implementation and execution on the Lean PLC in your day-to-day business. So while we work with early adopters, we also have to engage enabling functions in our design of the Lean PLC.

An example of this is finance. Remember that the Lean PLC is based on the notion of incremental investing. This is different from the traditional annual budgeting process that is used in most companies. To move from annual budgeting to incremental investing requires that new tools and ways of working be created. The finance team can help us solve questions such as:

- What are the financial thresholds or limits for each PLC stage?
- How do we release incremental funds and manage budgets?
- How do we account for the costs of innovation experiments?
- How do we reallocate budget from failing projects?

Another important function is legal and branding. A key part of using the Lean PLC is running experiments to test our ideas. Such experiment may require us to show customers minimum viable products or provide partial services. This has potential to damage the company's brand and create legal issues. These are real concerns that have to be dealt with, rather than being swept under the rug. We can work with legal and branding teams to create guidelines about what the product teams can or cannot do at each stage of the Lean PLC.

It is a similar scenario with the human resources function. In traditional HR, employees have defined roles and responsibilities. In contrast, the Lean PLC create a more flexible situation with regards to teams. In the early stages of the Lean PLC, there is an expectation that some projects will fail. The key HR question is what do we do with those teams? One thing we certainly do not want to happen is that people on failing projects lose their jobs. This would send the wrong message about innovation and creating a culture where people only work on 'safe' projects. As such, we need to work with HR to create flexible policies that allow people to move from different teams and projects within the company.

We could provide several more examples to help make the point (e.g. technology, sales and marketing). However, it is sufficient to note that when working to introduce the Lean PLC, the process touches every aspect of company life. The framework goes beyond product teams and affects those who manage or support the teams while they are doing their work. This is also why implementing generic standard frameworks does not work. You have to collaboratively design a bespoke version of the Lean PLC that works for your company and context.

Map your products

Of the many different approaches we have used to engage businesses, one of the most successful activities is having people map their own products to the six stages of Lean PLC. This is done in a fairly lightweight fashion where participants can answer just a few key questions to map their products. The exercise takes no longer than 10–15 minutes for most product managers and it involves completing a short questionnaire.

When we developed the tool, we iterated and experimented using Google Forms with controlled groups to ensure that the survey would be well understood, take as little time as possible for people to complete, but yield good data and learnings. Our main objective was for people to position their product on a Lean PLC stage, which would help them focus their efforts and increase their awareness of how to manage a product at a certain Lean PLC stage.

The mapping process also helps to socialise the Lean PLC within your organisation. When we first did the mapping work at Pearson, we were surprised by the level of engagement we received from senior leadership. Our initial goal was not to map the products from a portfolio perspective. However, the findings from the survey were so engaging because an interesting portfolio view emerged. Leaders began to ask interesting questions such as:

- How many products do we have at different stages of the Lean PLC?
- How many new ideas are coming through and how many fail?
- What is the average time and cost to develop new ideas?
- How many mature products do have in our portfolio?

☐ Do we have the right balance between the left side and the right side of the Lean PLC?

☐ How often do we retire failing products?

Our data were able to provide leadership with a view on innovation across the company in a meaningful way. Surfacing such portfolio data helped to illustrate the value of the Lean PLC at the strategic leadership level. This can create further buy-in for the process. In the end, our work accomplished more than just the mapping of products:

☐ We helped teams understand where their product was on its lifecycle journey and make decisions about what to do next.

☐ We were able to show leadership a portfolio view of the products and their lifecycle stages.

☐ Both of these helped to socialise the Lean PLC and get more support within the organisation.

Mapping your product to a PLC stage is the first step to managing the performance of your product and will provide a useful foundation for your growth performance management. Whether you have thousands of products or just one product, identifying the lifecycle stage for each individual product is a powerful sense-making exercise.

🔍 Mapping tips

☐ Before engaging your teams and organisation, we strongly advocate being clear with your intentions and reasoning. You must clearly state why you are doing the mapping exercise, how you will use the data and why their open and honest participation is important.

☐ Who should be responsible for submitting the product data? We advocate working with your dedicated product managers. If you do not have product managers, we would suggest identifying individuals who are making decisions on product design and delivery who also have oversight on the business model and growth projections.

☐ If you are in an enterprise with many products, it's advisable to agree on the definition of a product or service before mapping a Lean PLC stage. This consistent definition will need to be established for a consistent portfolio view and consideration of the business models.

☐ If you are mapping a single product across different markets, it's key to consider that the product could be in different lifecycle stages in those different markets. As such, you may wish to track the product globally and at a market/geographic level. We recommend that you explicitly state this level of granularity before mapping products.

☐ In our work with Pearson, we encouraged participants to first indicate the Lean PLC stage they believed they were in, before answering the questions. This was a useful decision teaching moment as it highlighted key differences and misunderstandings that led to productive conversations with each of the product managers

To help you get started with mapping your own product portfolio to the Lean PLC stages, we have provided a simple survey at the end of this chapter. Feel free to adapt and change the questions to make them more relevant to your context.

Create tools and playbooks

For an organisational culture to work it needs artefacts, tools and practices. Without these practical things the concepts of lean thinking will remain abstract ideas. What makes traditional management practices difficult to change is that they are embedded within companies via the tools and processes that companies use. Annual budgeting and forecasting are not just concepts or ideas in accounting books. These concepts are embedded in companies via the enterprise software, spreadsheets and business case templates that are a common in most companies.

For the Lean PLC to take hold within your company you need to develop similar a similar tool box. This book is based on a playbook that was initially created for Pearson product teams. This playbook provided a step-by-step guide to product teams on how to use the Lean PLC in their work. Like this book, it was organised by the Lean PLC stages and provided tools that the teams could use at each stage of the lifecycle. This playbook was the product teams' guide on how to do the right things at the right time.

At the end of every chapter in this book, we have shared with your tools for making incremental investment decisions. While teams can do the right things at the right time, this will not work if they are led by managers who ask the wrong questions at the wrong time. The traditional business case does not work as a tool for making decisions within the Lean PLC. As such, we created new tools that teams can use to make investment requests and managers can use to make decision. These tools are based on the Lean PLC stages and ask the right questions for each stage.

It is important that you develop similar tools for your organisation. Beyond playbooks and investment requests you will need tools for business model design, managing and tracking experiments, building prototypes and minimum viable products, talking to customers, running usability tests and agile product development. This toolbox will help bring the Lean PLC to life in your company.

Establish investment boards

Now that you have designed your bespoke framework and tools, it is time to create investment boards. The investment board is a small cross-functional team of leaders that make investment decisions on products using the Lean PLC. Companies can choose the level within the organisation at which they want their investment boards to sit. However, it is key that decision making happens as close to the product portfolio as possible. It is also important that the investment board has power over the relevant budget, so that it can make meaningful decisions. If necessary, an upper limit of investment decision making by the board can be established, with escalation points to senior leaders for larger investment decisions.

We recommend cross-functional representation at the investment board so that the discussions are informed by all the functions that are relevant to the success of the product (e.g. finance, sales, marketing, product). An investment board should be a small team with not more than 6–8 members. This allows the board to have meaningful conversations and make quick decisions. The board should also have a regular meeting cadence so that it does not become a bottleneck for teams that require decisions to move forward.

The investment board will make its decisions using the Lean PLC framework. As such, it is important that members of the board understand how the Lean PLC works. For each decision they make, they will have to consider the lifecycle stage of each product. This ensures that they are asking the right question and that their decisions are made using the right criteria. Below are some guidelines that companies can use to set up their investment boards. Feel free to adapt this guidance to the context of your company.

- The purpose of the investment board is to approve product investment requests on the left side of the Lean PLC (i.e. Idea, Explore, Validate), and to review product performance on the right side (Grow, Sustain, Retire).
- The board will manage individual products through the Lean PLC ensuring alignment to strategy, using Lean PLC tools, templates and stage-gate criteria to make and support investment decision making.
- The goal and remit of the board is to reduce the risk of wasted investment using incremental and evidence-based decision-making funding based on real data from the market.
- Members of the board should have regular and consistent attendance to board meetings as often as required (e.g. weekly, fortnightly or monthly).
- The board will provide transparency to teams with regard to decision-making criteria by providing visibility to meeting minutes and investment decisions.
- The board's standing membership will be 6–8 people from across key disciplines such as Business Unit Head, Product, Finance, Technology, Legal and HR as required.
- Guest members will be invited to attend if their input is relevant for making a decision on the specific request from a product team.
- A project manager will be needed for each council to manage logistics, prepare agendas, capture decision and distribute minutes of the meeting.

The first exercise of mapping your products to Lean PLC stages will provide you with a starting point to learn more about your products and condition of your portfolio. The power of the PLC is in the ongoing management of the products. Knowing which stage your products are in, reveals the key questions you need to ask and what indicators are available or need to be provided. If you have a lot of products to manage, you may need to consider establishing a small team to provide and manage a company-wide system. This can be lightweight and should look to integrate with existing systems where existing data can be pulled.

Whether you have 1 or 1000 products to manage though, what's important is that you actively manage the products in context to their Lean PLC stage. If you are a startup, you might want to consider setting up a core team to evaluate the product's performance on a regular cadence with key goals and data points. If you have investment partners, these partners may provide some of the objective performance feedback already. Adding the Lean PLC criteria should help meet their needs and increase their confidence in your startup. If you are a larger organisation, we would advocate establishing cross-functional product councils or product investment boards who act as objective reviewers of a product's performance. The Lean PLC stage governance should provide these product investment boards with the performance criteria needed to manage the products and portfolio from an investment perspective.

8.2 Build communities of practice

Introducing the Lean PLC into your organisation can represent a significant cultural change to existing practices. Any company will have strong business habits that people go to whenever they are working on products or making decisions. To change these habits and gain traction with Lean PLC adoption, you will need to create strong communities of practice. These communities will be safe spaces where employees at all levels can share learnings in a collaborative and open environment. Each community should look to celebrate where the teams have learned useful lessons and share ideas, practices and techniques.

As you begin your work, you can look for already existing communities whether inside and outside your organisation. An example of a range of communities that you may be may be able to connect with include: Agile Development, Lean Startup, Design Thinking, Branding, Marketing, Software Programming, User Experience and Customer Support. Each of these communities will likely have a very active and passionate membership. Connecting your company employees with these groups will help make them feel part of something bigger than just their company.

The valuable knowledge acquired through communities will go a long way to reduce waste and assumptions. Learning from others' mistakes is a great and cheap way to acquire wisdom. For instance, a product team looking to pitch their progress from Explore to Validate to a product council, may benefit from working with a marketing or pricing group experimentation beforehand. These savings vastly outweigh the cost of supporting communities.

In our work with Pearson, we created two types of communities. First came our Lean PLC coaches who we trained to provide support for product teams within the organisation. These coaches went through a two-day workshop and six-week hands-on practical training as coaches. After this they became members of a coaches community that supported each other with tools and practices. We also created an informal network of 19 enabling functions. We identified key champions within each function and worked with them to design Lean PLC tools and processes. Our champions and coaches were those individuals who were passionate about their craft and wanted to share knowledge. This helped product teams connect directly to knowledge experts to help answer questions quickly, reducing duplication and waste.

When setting up any communities of practice we recommend the following:

☐ Start small. You don't need to have a big group at the beginning. Instead, focus on passionate individuals who will participate in the community in a consistent fashion. As the group gains traction, you will find that more people will begin to attend.

☐ We strongly advocate a structured and explicit stance on diversity. If you're looking to innovate, a key ingredient is without doubt diversity. Taking a stance on diversity to support your communities of practice doesn't have to be difficult. As a good place to start we recommend engaging with the http://diversitycharter.org, where you will find countless resources to help structure your communities of practice.

☐ An important point to consider is that although you may develop internal networks first to initiate your community, it's recommended that your internal community connects with external communities.

☐ When connecting with external communities, we recommend avoiding communities who are preaching that their solution trumps all others or that they are smarter than others. These groups tend to be emotive and personally driven as opposed to being on the quest for knowledge and a better understanding.

PRODUCT PORTFOLIO MAPPING
A benchmarking survey

Tell us about your product.

Product name:

Product owner:

Business unit:

Briefly state your product's value proposition:

In which Lean PLC phase is your product today?

☐ Idea

☐ Explore

☐ Validate

☐ Grow

☐ Sustain

☐ Retire

Please explain your choice.

What is the main activity you are engaged in with your product right now?

☐ We are exploring customer needs and problems in context and we haven't yet started building the solution.

☐ We have started building the solution as a minimum viable version or a prototype to test with customers.

☐ We have launched the solution into the market and we have paying customers.

☐ We are growing customer numbers, revenues and profits every year.

☐ The solution has been in market for a few years and we have stable customer numbers, revenues and profits.

☐ The solution has been in market for a few years and we have declining customer numbers, revenues and profits.

What has been the approximate revenue from your product over the last three years, beginning with the latest year?

☐ Year 1: _____

☐ Year 2: _____

☐ Year 3: _____

☐ Not yet launched: _____

How do you envisage your revenue changing over the next year?

☐ Large increase (greater than 50 per cent)

☐ Increase (less than 50 per cent)

☐ Stable

☐ Decline (less than 50 per cent)

☐ Large decline (greater than 50 per cent)

☐ Don't know

☐ N/A No revenue yet

What have been the approximate number of customers for your product over the last three years, beginning with the latest year?

☐ Year 1: _____

☐ Year 2: _____

☐ Year 3: _____

☐ Not yet launched: _____

How do you envisage your customer numbers changing over the next year?

☐ Large increase (greater than 50 per cent)

☐ Increase (less than 50 per cent)

☐ Stable

☐ Decline (less than 50 per cent)

☐ Large decline (greater than 50 per cent)

☐ Don't know

☐ N/A No revenue yet

What have been the approximate product development cost for your product over the last three years, beginning with the latest year?

☐ Year 1: _____

☐ Year 2: _____

☐ Year 3: _____

☐ Not yet launched: _____

How do you envisage your product development costs changing over the next year?

☐ Large increase (greater than 50 per cent)

☐ Increase (less than 50 per cent)

☐ Stable

☐ Decline (less than 50 per cent)

☐ Large decline (greater than 50 per cent)

☐ Don't know

☐ N/A No revenue yet

What has been the approximate team size from your product over the last three years, beginning with the latest year?

☐ Year 1: _____

☐ Year 2: _____

☐ Year 3: _____

☐ Not yet launched: _____

How do you envisage your team size changing over the next year?

☐ Large increase (greater than 50 per cent)

☐ Increase (less than 50 per cent)

☐ Stable

☐ Decline (less than 50 per cent)

☐ Large decline (greater than 50 per cent)

☐ Don't know

☐ N/A No revenue yet

In which markets is your product currently offered?

Are you looking to expand to more markets in the next year? If yes, where?

*To review the responses to this survey and place a product or service in its Lean PLC stage, please use the descriptions of the Lean PLC stages presented in the Introduction chapter of this book. Please also note that you can adapt this benchmark survey to the needs and context of your company.

Case studies – applying the Lean PLC

As with most things, context is key. There are many ways, shapes and forms you can begin to apply the Lean PLC to your product or portfolio of products. You can apply the practices alone for a given phase, or roll out a company-wide initiative where the Lean PLC supports your investment governance decisions and performance goals.

Before applying the Lean PLC, it's important to know that behind any framework or model there sits the culture of your organisation. This culture is that of smart, capable intelligent people, who work together every day, collaborating to achieve successful goals. However you decide to apply the Lean PLC, we strongly advise you to take people with you on the journey and be open to new ideas and opinions. You are free to extend and contextualise the Lean PLC to suit your needs.

To help you consider the different ways the Lean PLC can be applied, this chapter consists of several case studies from companies who have applied or are on the journey of applying practices that we have advocated. We hope the differences and similarities provide a foundation to help you take the next steps.

SDL – Rolling out the Lean Product Lifecycle

URL	http://www.sdl.com	
Primary business domain	Distribution & Size	Lean PLC focus
Language translation technology, services and content management	Global customer base, 54 locations worldwide. >3700 Employees	☐ Product and Portfolio Management ☐ End-of Life Management ☐ Innovation Accounting

Known as a global innovator in language services, translation technology and content management solutions, SDL has helped its worldwide customers deliver nuanced digital experiences for the past 25 years.

SDL is uniquely positioned in the marketplace by its combination of technology and services. SDL localises more than 100 million words by human translators and 20 billion words by machine translation per month. In addition, its translation technology and web and documentation content management solutions enable organisations to manage and deliver multilingual content across channels. SDL powers digital experiences for 78 of the top 100 global brands, diversified over a large number of industries.

In 2016, SDL appointed a new CEO Adolfo Hernandez with a supporting executive team, who set a strategy to expand their language and technology footprint. This is supported by an ambitious growth strategy to become the best in class $1bn business and tactically embark on a journey to become more agile and scalable.

As part of a global multi-year business transformation program, SDL initially chose to implement the Scaled Agile Framework® (SAFe®) to meet its goals. This consisted of delivering value while reducing time to market, improving quality, and strengthening alignment and collaboration across geographically distributed multi-disciplinary teams and functions with reduced cycle time.

With the initial focus on implementing SAFe® within the product management and DevOps community, SDL quickly realised the company would benefit from a PLC management approach. This required extending the SAFe® framework predominantly in the areas of portfolio management, end-of life management and innovation accounting. the goal was to globally improve the management of its innovation pipeline, work in progress, and new product market launches into existing and new markets while increasing transparency of the overall global product portfolio and improving how it recognises and works on the retirement of products and services.

Geert Wirtjes, VP of business transformation at SDL, is responsible for leading the transformation to apply the Lean PLC. This includes the product portfolio and organisation as a whole.

The journey of applying the Lean Product Lifecycle at SDL

Led by Geert Wirtjes, a small dedicated team was established at SDL supported by CA Technologies. This team acts as a centre of excellence and is fully backed by an experienced executive management team.

One of the first tasks for this team was to baseline and prioritise the changes needed based on the value of each proposed transformative change. This took a pragmatic approach that would address Portfolio and Program Management, align the SAFe® implementation program to the Lean PLC, analyse and prioritise the continuous integration and delivery capabilities and assess the related enabling infrastructure.

To support this initiative, the team decided to pledge to 'Eat Their Own Breakfast' by using the Lean PLC phases and methodology to deliver the transformation program itself. This resulted in the decision to consider submitting ideas, then exploring and validating the components that needed to be addressed, including aligning the budget cycles, growing the teams and developing skills over time as the implementation takes place.

When researching how the company could adopt a Venture Capital model to fund new ideas and product portfolio, SDL identified that it needed a stronger lean and agile culture across the business. Capturing this need, the transformation team embarked on a series of top-down alignment sessions with all relevant functions, and starting with the executive team. This further extended across finance, legal, marketing, product management, sales, business development, support, operations, DevOps and more.

The team used the Lean PLC framework and material to engage the various functions, resulting in the creation of a minimal viable product (MVP) approach to change management across existing frameworks and operations.

Following detailed designs, it was found that the Lean PLC and SAFe were compatible for SDL at an operational level. Breaking down the Lean PLC stages and goals, the team identified key activities for functions, domains and job profiles at each lifecycle stage. It was beneficial to identify roles and responsibilities clearly through the creation of a RACI (responsible, accountable, consulted, and informed) referenced against each Lifecycle phase and mapped to existing SDL roles and the newer SAFe roles. This identified the roles that would be needed for different phases and possible skill gaps or confusion, which showed the ways in which the organisation was weighed between innovation and execution.

Supporting this new operational model, SDL decided to customise Rally software to work at a Lifecycle phase level, allowing for a single source for a global product portfolio view within the lens of each Lifecycle phase. This also acts as a way to capture and nurture new ideas through each Lifecycle stage, becoming a living knowledge base and learning hub for existing and future innovations. As ideas make it through each of the subsequent Lifecycle phases, their progress is recorded, whether they stop, pivot or continue. Ideas which make it through Explore, Validate and Grow phases, then evolve into the extended SAFe operational framework for execution.

Through a series of further introduction and experimental workshops and engagements across SDL, the application and alignment of SAFe and the Lean PLC began to be adopted. Supporting the rollout, a global training program was initiated which included Standard SAFe training (Leading SAFe®, SAFe® for Teams), on the job coaching and functional PLC training for each phase. Within just a few months, the first product council and quarterly review boards were established. As these teams formed, this new structured approach started to show the availability of a global inventory of innovation pipeline for each of the lines of businesses.

Soon after implementing the steps above across SDL, Geert and his team observed and recorded the following benefits.

1. Almost immediate results in more collaborative alignment across functions supported by simply channelling / focusing activities and having alignment discussions.

2. Consolidation of input channels into the Innovation and R&D Pipeline, with improved transparency and feedback loops on the status of 'ideas and suggestions' coming in.

3. Harmonisation over Product Lines.

4. The PLC applied using Rally has become an internal innovation management platform. All this now leads into a central and global managed portfolio Kanban board.

5. The program and application of the PLC has highlighted operational inconsistencies. This has identified initial efficiency gains and improved flow, by re-organising roles and responsibilities holistically on an operational level.

6. Reduced risk and liability due to early involvement of and alignment with legal (patent management), data privacy, security in earliest stages of software development.

7. Wake up call to 'Get out of the building' and introduction of an 'evidence-based' mind-set challenged through PLC governance and QBRs (quarterly business review) of our executive team.

8. Standardised ways of working increased transparency on portfolio level and improved capabilities for rationalised/priority-based decision making.

9. Repositioning of its UX (user experience) competency to fully support the earliest stages of the PLC including training and support on lean design principles.

10. Harmonisation of job profiles and embedding into HR systems and approach.

In reflection of where SDL are on their journey of adopting and applying the Lean PLC, Geert Wirtjes states, 'Although we have just embarked on a multi-year program, our transformation journey could not have started more efficiently and effectively. It has been a tremendous learning experience for all involved with almost immediate positive effect on productivity, time to market, quality and job satisfaction. We will continue to deliver significant value to all our stakeholders and customers by universally adopting the practices and behaviours outlined in our progress above, and continue to find new ways to create innovative, profitable products, whilst maximising customer value.'

The next steps for SDL involve the further expansion of the product councils to cover all business lines. This would complete the holistic view of the product portfolio, managing innovation and execution within each lifecycle stage.

Insight Software

URL	http://www.gohubble.com	
Primary business domain	**Distribution & size**	**Lean PLC focus**
Corporate Performance Management – JDE Edwards, Oracle EBS	Global customer base operating in over 10 regions. <200 employees	□ Product management □ Market growth □ Lean & agile development □ Innovation

Insightsoftware.com formed in 2000 originally specialising in highly performant financial reporting over JD Edwards. The founders realised a trend that many financial professionals had regarding not being able to extract the critical data and reports they needed on an ad-hoc, timely basis from JD Edwards without having to involve IT. This problem led to for the development of Insight Reporting – a much-loved product and recognised tool for financial analysts and controllers.

Keeping the customer needs at the forefront of research and development, Insight Software expanded their solution offering in 2015 with the introductions of Hubble, a real-time, comprehensive financial performance management solution. Recognising the need to offer and distribute data across organisations in a meaningful way, in 2016 Insight Software acquired Decision Point as part of Antivia.com; an information delivery platform. This with many other innovations formed part of their growth strategy to distribute the right data at the right time to the right people, supporting them to make better decisions.

With an expanded and competitive solution offering alongside innovative global partnerships, insightsoftware.com successfully serves over 30,000 users globally in industries such manufacturing, distribution, oil & gas, retail, real estate, services, pharmaceuticals and more. Insightsoftware.com now has a global reach with key partners in many regions and is a rapidly evolving its product offering and capabilities.

Using the Lean PLC to maximise learning and improve product outcomes

Putting product development at the forefront of organisational strategy, in 2015 insightsoftware.com began to re-shape internally. In order to meet the company's growth ambitions and goals and be able to scale the product offering into a broader enterprise market, the organisation brought on new additions to the senior leadership team within the areas or product, engineering, services and marketing. These new additions started to call upon some lightweight practices and principles offered by the Lean PLC at a product level at the relevant product phases.

Holistically the Lean PLC as a complete model could be seen in its entirety as a framework for larger enterprises if you consider many products and each lifecycle stage. However, the Lean PLC can be utilised as much or as little as possible and extended where appropriate to better suit your business and product context. Like any framework, it's value is the contextual application to your needs. Being an SME, Hubble did not require portfolio management functions or the product council governance that the Lean PLC can offer. As a result, the Lean PLC was introduced much more at a practice level for product management and product development, evolving depth over time at the pace of culture adoption.

To take action on the opportunity, the products were mapped and are managed according to their lifecycle stage. The governance is lightweight and focused, meeting the product teams where they are initially and building up from there. The Lean PLC product council is effectively the board of directors who meet on a monthly basis to discuss progress, growth and investment. Although budgets are used, they adapt rapidly where needs surface to position the teams and products for success. The benefit in doing so is that new opportunities can be capitalised on or where new capabilities can expand, pivot or stop to meet the changing customer needs. This provides an adaptive management response allowing the business to respond quicker to change, but it's not all about expansion. Like any business investment is not infinite, so this way of working allows the company to also make decisions on what are the priorities and what are not.

In order to improve the customer development and idea validation capabilities, Hubble have explicitly set up their product marketing team to be the leaders in Explore phase; running rapid controlled experiments and customer engagements with the support of the sales and support teams. During Explore they experiment using hypothesis-driven design, business modelling and customer development practices at a rapid pace and engage all functions across the business to support them. New ideas are put through a timed programme which involves the relevant functions of the business including the global sales team.

When unequivocal evidence starts to emerge from the Explore stage containing evidence through customer engagement, only then do they begin to engage product development teams to start considering the points in Validate. This is usually aggressively time-boxed to just a few weeks or months depending on the complexity of the experiments and engagement resource availability. The learnings and outcomes from these exercises then either naturally make it into the product roadmap with the first stage of supporting evidence, will require a pivot for the team to explore new avenues or can surface additional investment opportunities to the investment board which stretch beyond existing commitments.

Pioneering some of these Lean product practices, Kevin Josling, Hubble's product marketing manager stated:

> *We have been using the lean experiment canvas to help us structure our approach to launching a new product idea, and it is really paying dividends. The framework has made us think more deeply about the assumptions that underpin the idea, and devising the experiments gives us an opportunity to really explore those assumptions – a couple of times we eventually even changed the definition of the assumption at this phase simply because the conversation we had about designing the experiment surfaced an even more critical assumption that we had overlooked.*

> *Similarly, the approach paid dividends in time of speed of learning and reducing costs: we now take a minimal approach to designing tests, and rather than building products we now gravitate to low-friction solutions – testing out ideas through prototypes, surveys, or even putting out questions through our social media channels to validate the central assumption, and giving us time to iterate another test to dig deeper into the assumption.*

Following any successful outcomes from Explore, the product management team at Hubble, directly apply some of the practices and some of the tools shared in the Validate and Grow phases to accelerate product growth and maturity. This includes much of the roadmap, experimental hypotheses structure using evidence as a multiplier for confidence and priority. All ideas are scored and prioritised on a continuous basis. Where new ideas are proposed from anyone and anywhere in the business, they must support the idea with evidence which can come in any shape or form.

Allowing everyone to see, understand and contribute to the product roadmap can be challenging, particularly in a global company across many time zones. Overcoming this, the Hubble product management team lead by Naill McLean, researched and introduced new platforms such as Aha.io and found a practical way of sharing progress across the company to keep everyone aligned. Hubble has managed to deliver the product roadmap across the organisation in a way that enables progression while providing complete visibility at all times. Everyone from the board members to the global sales teams can see a single source of the truth at any time using the real-time product roadmap.

The management team review the roadmap at least once a month and have structured governance in place emulating the Lean PLC. This includes every function including delivery, QA, engineering, support, product marketing and product management having operational KPIs shared with everyone.

Though the roadmap is expressed to the organisation in its current state, a collaborative innovation portal exists to flow ideas into the development team. This area is referred to as, Hubble Ideas. Hubble Ideas allows employees within the organisations to contribute to the roadmap. Once submitted, each idea is evaluated and responded to on a weekly cycle. There is a scoring guide in place which ranks the idea against strategic goals and customer needs. This is then balanced against the time to implement and company priorities. A key multiplier is evidence. So when ideas are submitted, some direct evidence of demand, need or opportunity is required in a lightweight format, which sometimes just involves customer emails. Such evidence can multiply over time as others vote and contribute to the idea. Ideas that are successful are then positioned for exploration and then mapped to the roadmap accordingly.

Alongside Aha.io and behind the product roadmap the distributed development teams use JIRA, to utilise by the development and services teams. This provides the organisation live two-way integration into the roadmap in Aha.io which is useful at presenting data at the right level. This abstraction also means that the development teams' workflows are not needed to communicate progress to the rest of the company, this is simplified at a higher level. The many development teams don't apply any large-scale agile frameworks, but simply have a range of workflows which each team evolves mainly utilising Kanban and Scrum. This all maps back to the product roadmap in real-time aligned to key initiatives.

Communities of practices have formed and continue to evolve supporting the product, skills and goals. Teams and functions connect and work together on a regular basis with product teams structured into functional communities by their craft. Bringing the organisation together, Hubble has four knowledge councils and review boards operating on a three-week cycle which bring together all business functions including sales, marketing, product, engineering and more. Here they share ideas and contribute suggestions on how to progress forward together and the councils are empowered to act on the information and make decisions immediately.

To maximise learning, a culture has also emerged where a 'five whys' exercise is carried out following any event that didn't go to plan or did where there was known risk; covering service deployments, upgrades, development obstacles and more. This has significantly reduced the number of issues that could be repeated. Supporting this, the management team granted all employees individual unlimited account for Udemy for Business to remove request barriers and line managers have to explicitly support and manage at least a 10 per cent reserved time for personal development.

The next step for the Hubble team is to evolve their management structures to be more explicitly aligned to the Lean PLC governance, providing a real-time view connected to the company's growth goals and strategy.

Hubble recently created a separate team which sits outside the core business, managed by one of the original founders Paul Yarwood. This team, free from the core culture, explores new potential adjacent and transformational applications for the Hubble products and is not measured by the same governance as the core business. When new opportunities surface, they will be pitched back to the business to take through Explore, Validate and Grow.

Index

Page numbers followed by *f* indicate figures; those followed by *t* indicate tables.

To the future and beyond

To the current Product Lifecycle Team at Pearson led by Amy Burmeister. To Gabe Gloege, David Alick, Carol Hill, Yazad Cooper and Paige Patunas. Thank you for picking up the baton and doing a fantastic job of reimagining, improving and embedding the Lean PLC across the company. We wish you the best successes for the future and beyond.